£15

**AMAZON
TO
IVANHOE**

AMAZON. Note the canvas portions to her bridge and the half-shields of her 4.7" guns.

(National Maritime Museum N9348)

AMAZON TO IVANHOE

BRITISH STANDARD DESTROYERS OF THE 1930s

BY

JOHN ENGLISH

CONTENTS

Introduction	5
Background to the development of the Prototypes	7
Torpedo Boat Destroyers (dated 11.23)	7
The Tendering of the Prototypes	8
The Assessment of the Designs	9
Report of Trials and Long Distance Endurance Test	10
The 1927 Programme, CODRINGTON and the ACASTAs	13
SAGUENAY and SKEENA (R.C.N.)	25
The 1928 Programme, KEITH and the BEAGLEs	29
Proposals by Yarrow and Thornycroft for a two boiler design	30
The 1929 Programme, KEMPENFELT and the CRUSADERs	43
Proposals for Leaders armed with 6" guns	44
The 1930 Programme, DUNCAN and the DEFENDERs	51
The 1931 Programme, EXMOUTH and the ECLIPSES	62
The Minelaying Destroyers	63
The 1932 Programme, FAULKNOR and the FEARLESSes	75
The 1933 Programme, GRENVILLE and the GREYHOUNDs	88
Proposed Modifications to the vessels' A.A. Armament	89
The 1934 Programme, HARDY and the HEROs	102
The 1935 Programme, INGLEFIELD and the INTREPIDs	114
The Brazilian 'H' Class	127
The Turkish I's	135
Appendix I: Deployment	137
Appendix II: Armament changes to the A/I Destroyers on Conversion to Anti-Submarine Escorts	141
Appendix III: Armament fit of FAME on Transfer to the Dominican Republic	142
Index	143
Acknowledgements	144
Photographs	144
World Ship Society Publications	144

Front cover: INTREPID while serving with the Home Fleet 3rd D.F. in 3.42

**Published by the World Ship Society,
28 Natland Road, Kendal, LA9 7LT, England**

ISBN 0 905617 64 9

© 1993 J.I. English, M.Sc. (Trans.), B.Sc. (Econ.)

INTRODUCTION

The A to I's were typical products of the Admiralty Ship design process, in that their design was based on the best destroyer design extant in 1923 — the Improved 'W' class destroyers of 1918.

This design was then modified in light of subsequent experience and with the addition of the newly developed ASDIC submarine detection equipment, wireless telegraphy, larger electrical generators, improved ventilation, better lagging and better headroom below deck. This all added to the complexity and cost of the vessels. The pom-pom short range anti-aircraft weapon was also to be fitted, but was still under development. Other additions to the two prototypes, constructed under the 1923 Estimates, were greater storage capacity for munitions and stores and improved washing facilities for their complements of 140 compared with the 127 of the Improved W's.

It was the Thornycroft prototype AMAZON which was developed for series production, after trials with the Yarrow-built AMBUSCADE and after designs from Hawthorn Leslie, Denny and White's had earlier been rejected.

A new Staff Requirement for destroyers was issued on 20.12.26 with the highest priority being given to the vessels' torpedo armament of MkIV 21" torpedo tubes and not their 4.7" guns. The need to withstand ocean weather conditions and to maintain fleet speed in moderate weather conditions were also important. Maximum speed was to be $31\frac{3}{4}$ knots in deep load condition.

The new vessels reflected the traditional role of the destroyer as a fleet escort, where its large torpedo armament was to be used to fend off attacks by enemy destroyers and scouting vessels and to attack enemy fleet-units with torpedoes. Little thought was given at this time to the use of the destroyer as a general purpose escort or to the increasing threat from air attack and the growth in the capabilities of the submarine, with its greater range, speed and diving depth.

However, when the A's entered service in 1930 as the primary development of the AMAZON they were roundly criticised by the sea commands as being too large and costly. Sea officers also preferred sustained sea speed in poor conditions to a high maximum speed.

Throughout their development the A to I's were subject to restraints in cost and better vessels could have been constructed for a moderate increase in costs. The whole group were weakened by being armed with 4.7" guns that had only 30° elevation, when other nations, especially the Japanese, had developed moderately effective dual-purpose weapons for their destroyers as early as 1928. The severe reductions in British ordnance capacity, especially in design teams, meant that there was never an effective short-range anti-aircraft weapon in British service throughout the 1930s. It was only with the adoption of the 20mm Oerlikon in 1938 and later the 40mm Bofors, that this deficiency was remedied, but at the cost of nearly a score of A to I's being lost to air attack in the first three years of the war.

Throughout the early years of their development, there was prevarication on the need for a larger vessel for command and control purposes against the need of a homogenous group of vessels. CODRINGTON and EXMOUTH were early leaders, whilst KEITH, KEMPENFELT and DUNCAN were of the destroyer form modified to meet command duties, with the suppression of a 4.7" gun and its replacement by increased accommodation for a Captain's staff. Later the 'E' to 'I' Flotillas had leaders, but later practice was to revert to a wholly destroyer design to save money.

The machinery and boilers of the vessels was conservative in design, but after teething problems with the A's proved to be generally reliable. ACHERON's high pressure boilers gave constant trouble and she was soon relegated to subsidiary duties. It was not until the 1936 Estimates and the 'J' class that higher boiler pressures were adopted, as was the two boiler arrangement, with its consequent reduction in silhouette. The unit system where boilers and machinery were grouped together, with consequent increases in survivability, was not adopted in the U.K. until 1943.

During the decade of development, repeated attempts were made to increase endurance, after the misguided decision with the A class to reduce bunker capacity when it was realised that the propulsive efficiency of the prototypes was greater than estimated and that oil-fuel bunkers could consequently be reduced. Attempts were made with the C's (1929) and E's (1931) to improve range, when oil-storage was increased to 445 tons — an increase of 55 tons over the A's. The endurance of the vessels reached 5,530 miles at 15 knots with the G's of 1933. GIPSY of this group was fitted with an experimental Johnson boiler, which proved to be effective in service. The ten years of development also saw a considerable reduction in machinery weights, which was to be especially beneficial in later years, when the group became weight critical.

Considerable extra equipment was also fitted to vessels of the group — with ASDIC being fitted as standard after the E's. The vessels were also fitted with the twin speed destroyer sweep, whilst two E's and the I's were also fitted as minelayers. The number of depth charges was increased to 20 — a fraction of what it was to become standard during the war years and efforts were made to augment the close range high angle A.A. armament, by the replacement of the original 2 pounder pom-pom by a high angle 3" gun of World War I vintage between 1940/41. The low angle 4.7" guns were retained, only because a proposed 5.1" gun proved to be impractical in use and proposals to fit the G's with twin 4" and later with a 70° 4.7" in 1936 were abandoned as the G's were too far advanced to be altered. The need to expedite the construction of the H's and I's and the development of the "Tribals" and J/K's meant that further development of the A to I's was abandoned and all design effort concentrated on designs with more development potential. By 1936 weight problems were affecting the H/I's which were ballasted from commissioning.

The A to I's met the requirements of the British Staff — they were relatively cheap, reliable and plentiful — not too large, with an ability for sustained operations in heavy weather. They, however, were more suitable for the large fleet operations of the previous war, than the combined air, surface and submarine threats that they were to face during the second conflict. They were to prove woefully susceptible to air attack, but so were most of the destroyers built up to 1939.

Despite severe losses off Norway, North-west Europe and especially in the Mediterranean during 1940/41, the vessels served with distinction and held the ring until replacements on a large scale became available from 1942. The class was still to serve on fleet duties in Arctic waters and in the Indian Ocean until 1943. However, by this date the survivors had been transferred to escort duties on the North Atlantic where the vessels were modified to suit these new tactical conditions, when and where equipment and repair yard capacity was available (see Appendix 2 for details).

Alterations were many and varied, with the suppression of 1/2 4.7" guns, the addition of 3" guns, which were superseded at first by 20mm and then by 40mm weapons. Torpedo tubes were reduced in number and those fitted adapted to fire A/S torpedoes or the one-ton depth charge. Depth charge stowage was increased manyfold and new anti submarine weapons, such as the Hedgehog fitted. The electronic equipment fitted was to revolutionise the vessels' performance and such equipment included the fitting of various types of radar, radio-telephone equipment, direction finding equipment and improved ASDIC. That the vessels proved suitable for such a change in duties is a testament to the basic soundness of their design and construction, if not their equipment.

OTTAWA (II). Compare with AMAZON. She shows a mass of war modifications with 4.7" guns reduced to two, Hedgehog and Type 286 radar added, up-rated A.A., extra depth charges and one set of torpedo tubes removed.

BACKGROUND TO THE DEVELOPMENT OF THE PROTOTYPES

By 1923 the Royal Navy Destroyer flotillas had been run down from a strength of over 400 in 1918 to some 185 vessels of the M, R, S and V/W classes together with 15 leaders of the MARKSMAN, BROKE, SPENCER and SCOTT classes. All pre-war destroyers and a large number of the war-built M and R classes had been scrapped between 1921/23 for reasons of economy, material condition or under the conditions of the Washington Naval Treaty.

Of the vessels retained, the survivors of the M, R and S classes had no development potential because of their limited endurance so the first line destroyers consisted of the highly successful V, W and Improved W class destroyers and Leaders. A score of the vessels of the Improved W class and the Leaders of the BROKE and SPENCER classes, which survived the mass cancellations of 1918/19, had slowly been completed in the years up to and including 1924. By that year it had been nearly seven years since the last destroyer design had been developed and it was a necessity for the Admiralty to develop a new design that incorporated the technical changes that had occurred during this period. On 21.11.23 the Board approved its first programme of new construction since the war, consisting of eight 10,000 ton Cruisers, two Destroyers, three Patrol Submarines, one Submarine Depot Ship, one Destroyer Depot Ship, one Minelayer and two Yangtse River Gunboats.

The two destroyers, ultimately the prototypes AMAZON and AMBUSCADE, were the beginning of a line of development that was to total over 80 vessels for the Royal and other Navies. The Royal Navy's programme was completed in 1937 with the I Class, but eight near sisters being built in the U.K. were requisitioned during 9.39 and completed between 12.39-3.42.

The specification of the prototype had first been discussed at the Controller's conference of 13.11.23 which agreed the following basic characteristics:

Shaft Horse Power	40,000 to achieve 34 knots on full power trials.
Radius of Action	5,000 miles at 12 knots at 85% of fuel capacity and 1,200 miles at full speed.
Armament	To consist of 2 triple torpedo tube mountings; 4-4.7" guns with 150 rounds per gun; 2 Mark 'M' pom-poms. (Note that the torpedo armament predominated over the gun armament.)
Length of Vessel	310' between perpendiculars.
Other Equipment	To be fitted with ASDIC and Wireless.

The design was basically a repeat of the Improved 'W' Class which were generally the last destroyers to have been completed by the tendering companies.

Following this meeting the Director of Naval Construction (D.N.C.) drew up three sketch designs for the ships which were still classified as Torpedo Boat Destroyers.

TORPEDO BOAT DESTROYERS (DATED 11.23)

	No. 1	No. 2	No. 3
Length between perpendiculars	320'	330'	335'
Beam	33'	34'6"	35'3"
Deep load draft	13'3"	13'8"	13'8"
Deep load displacement	2,330 tons	2,540 tons	2,625 tons
Weight:			
	Tons	Tons	Tons
Hull	740	850	900
Machinery	650	650	650
Equipment	100	100	100
Armament	90	90	90
Max. Speed (Knots)	33.5	33.5	32.2
Endurance at 12 knots (Miles)	4,500	5,050	5,000
GM in deep load condition (ft)	1.43	1.58	1.84

In a note dated 10.12.23, the Engineer in Chief estimated that the machinery, less electrical apparatus, would be 512 tons for a two boiler arrangement and 529 tons for a three boiler arrangement.

Following discussions between the Controller and the DNC, further parameters of the design were agreed:
(i) The vessels were to be limited to 315' between perpendiculars.
(ii) The vessels were to be constructed in 'D' quality steel.
(iii) The equipment of the vessels was to include the same armament as the 'W' Class, ASDIC, Wireless Telegraphy, a Mark 'M' pom-pom.
(iv) The vessels were to have a 3 boiler arrangement on a displacement of 1,530 tons and a trial speed of 34 knots.
(v) In addition, the vessels were to include a total of 24.5 tons of additional weights that were not included in the Improved 'W' Class vessels. The items making up this total were two generators (36 kW) (3 tons), larger magazine (0.5 tons), improved ventilation (1 ton), ASDIC (12 tons), additional lagging (1.5 tons), additional awning stanchions (1.0 ton), wireless telegraphy room (1.5 tons), additions to forecastle (3.0 tons), larger bridge (0.5 tons) and additional deck stiffening (0.5 tons).

The fighting qualities of the proposed vessels were to be improved by the storage of an extra 50 rounds per gun, and by the fitting of better fire control equipment. After early trials, it was reported that:
"The installation as a whole has justified itself in these trials in that it gives considerably better results than those possible with existing fire control equipment in destroyers."

The cost of AMAZON's equipment was £17,400 compared with that installed in a repeat 'W' Class of £4,625.

Habitability in the vessels was improved for the crew of 140 (compared with 127 of the 'W's) by the installation of separate washing facilities for C.P.O.'s, P.O.'s and ratings and increasing deck height by 2" to 8'.

THE TENDERING OF THE PROTOTYPES

On 26.11.23 five of the foremost destroyer builders were asked to tender — Denny, Hawthorn Leslie, Thornycroft, J. S. White and Yarrow. After the plans for the designs were discussed between the Admiralty and the potential builders, tenders for 2 prototypes were finally invited on 29.2.24. The tenders were to be received by the Admiralty by 11.4.24.

Thornycroft provided two sketch designs.
Design 'A' 310' (PP) which followed closely to the sketch arrangements.
Design 'B' 305' (PP) with a single boiler forward as in WISHART.

The details of the designs are given in Table 1.

Table 1. LEGEND OF PARTICULARS OF TENDER DESIGNS COMPARED WITH ADMIRALTY SPECIFICATION

Characteristic	DNC Proposals (12/23)	Thornycroft Proposals I	Thornycroft Proposals II	Yarrow	Denny	J. S. White
Length (PP)	315'	310'	305'	307'	310'	310'
Length (OA)	328'	323'	317¾'	322'	323'	323'
Breadth (Extreme)	32¾'	31½'	31½'	31'	31½'	31½'
Deep Draft (Mean)	12¼'	13½' (aft)	13⅙' (aft)	10' 1,600 tons	—	9' 1,622 tons
Shaft Horse Power	33,000	35,000	32,000	32,000	33,000	33,000
Speed (load)	34 knots	36	35	37	35	35
Oil Fuel (tons)	710	470	450	275*	320*	385
Complement	140 men	—	—	—	—	—
Weight (all tons)						
General Equipment	84	127	125	170	30†	—
Armament	114‡	—	—	—	—	—
Machinery	520	486	485	474	495	495
Hull	700	497	478	582	610	610
Margin	2	—	—	—	—	—
Displacement	1,420 tons	1,110	1,088	1,226	1,135	1,105

Oil Fuel and Reserve Feed Water to be added to give deep-load displacement.

‡ Includes Armament for D.N.C. proposals, but excludes armament for the remainder.
* To give guaranteed radius of action.
† Given in Ships Cover, but seems to be very low compared with other general equipment totals.
Note: No details of the Hawthorn Leslie design have survived in the Cover.

THE ASSESSMENT OF THE DESIGNS

The D.N.C. commented on the tendering process on 23.5.24: "It was desired in these two destroyers, that we should obtain the experience of the designing firms, with a view to making some advance in destroyer practice, principally in connection with the machinery arrangements, so as to obtain good full speed and in particular, good endurance performance at an economical speed".

Generally, the Thornycroft 'A', Hawthorn Leslie, Denny and White designs were close to the Admiralty pattern, whilst the Thornycroft 'B' and Yarrow designs differed in their machinery arrangements and the design of the after-end of the forecastle.

The Hawthorn Leslie design was criticised as having heavier scantlings than was thought necessary, the vessel appeared to be 60 tons overweight, living accommodation was too small and machinery space too large and the company could only guarantee 34 knots.

The Denny design was criticised for being 50 tons overweight. Whilst the J. S. White design had a guarantee of 35 knots, its draft was 4" greater than specified.

The Thornycroft 'B' design was preferred to the 'A' design. The vessel had the same displacement as the Admiralty design. Thornycrofts had not cut the scantlings as Yarrows had done and the design was considered satisfactory both for stress and stability. The Yarrow design was criticised for its light scantlings, which were considered to be below what was desirable.

The order of preference of the tendered designs was:
(i)   Thornycroft 'B' and Yarrow
(ii) Thornycroft 'A'
(iii) Denny, Hawthorn Leslie and White designs

On 12.6.24, the tenders of the Thornycroft 'B' and Yarrow designs were accepted by telegram.

Estimated Costs:

		Yarrow	*Thornycroft 'B'*
(1)	Hull	£101,000	£82,490
(2)	Main and auxiliary machinery	£164,000	£133,300
(3)	Other equipment and spare guns	£9,000	£9,700
		£274,000	£225,490

The far lower Thornycroft tender could either reflect building efficiency or the desire to get the order.

By 1.7.24 the Yarrow vessel had been allocated the name AMBUSCADE and Thornycroft's vessel had been allocated AMAZON. Some three weeks later, on 23.7.24, Thornycrofts accepted a reduction in the beam of the proposed vessel.

During the summer and autumn of 1924, Thornycrofts considered the design of AMAZON and proposed to modify it in the following ways:
(i)   Engine room to be lengthened by 2 frame spaces.
(ii) Boiler room to be lengthened by 1 frame space.
(iii) After magazine and shell room to be reduced by 1 frame space.
(iv) Stern compartment to be shortened by 1 frame space.
(v)   Oil tankage to be increased by 470 tons.
(vi) The machinery to have the same general arrangement as the D.N.C. design.

The preliminary estimated cost of the revised design was calculated at £273,000. However, on 6.1.25, the Admiralty agreed to place an order in lieu of the original order, at a price of £269,300 of which the make-up was:

	£
Hull	90,490
Galvanising	3,660
Inclining	150
Machinery	159,700
Spare guns	11,000
Auxiliary Turbines	4,300
	269,300

The vessel was to have the same radius of action as the previous design and to be completed on the original date (21 months from 7/24).

	THORNYCROFT	
	ORIGINAL DESIGN DATED 4/24	REVISED DESIGN DATED 11/24
Length (O.A.) ft	317' 9"	323'
Length (W.L.) ft	314'	319'
Length (P.P.) ft	306' 9"	311' 9"
Beam ft	31' 3"	31' 2"
Depth ft	19'	19' 6"
Draft ft	13' 5"	14' 3"
Speed (knots)	35	37
SHP designed	31,500	36,000
SHP max.	34,500	39,500
Weights (Tons)		
Hull	478	504
Fittings	125	131
Machinery	485	525
Load	170	170
Reserve Feed Water	20	20
Oil for 3 hours	50	60
Displacement at Mid trial	1,328 tons	1,410 tons
Price	£233,600	£273,000 (Amended)

NO SIMILAR DATA SURVIVES FOR THE YARROW DESIGN

The full speed trials of AMBUSCADE and AMAZON took place on 2.3.27 and 12.3.27 respectively. Mean speeds achieved were 36.88 knots for AMBUSCADE on a displacement of 1357 tons and 37.47 knots for AMAZON on 1515 tons.

REPORT ON TRIALS AND LONG DISTANCE ENDURANCE TEST

As the vessels contained so many new features and much untried equipment, it was decided to test this equipment over a period of time and also assess the efficiency of the new machinery and boilers.

A long distance trial from Portsmouth to Pernambuco was undertaken during 4—8.28. The vessels had to maintain a constant number of revolutions. "This proviso resulted in a marked drop in speed when head winds were encountered, but in the tropics with a following wind, there was not an increase in speed owing to the warm sea water causing a fall in vacuum in the condensers."

The C.O. of AMBUSCADE reported that the vessel suffered vibration when the engines were warmed-up quickly, at between 25-28 knots and again at speeds of over 35 knots.

The C.O. of AMAZON was critical of the super-heated machinery for the following reasons:
(i) The higher temperatures in the machinery spaces reduced the energy of engine room complement.
(ii) The machinery was complicated.
(iii) Maintenance was more difficult.
(iv) The boilers were more difficult to clean and needed more time for maintenance.

It was appreciated that superheated machinery was a recent innovation and as such was subject to fear by conservatives.

Overall AMAZON had greater fuel efficiency than the AMBUSCADE. However, as the ships got lighter instead of the vessel speed increasing, the opposite was true. The report continues

"The loss of efficiency due to the fall in vacuum and greatly increased windage as the ship got lighter more than counter-balanced any increase of speed, which might have been obtained by lightening the ship. The vessel became more lively and required more helm and the consequent loss of speed resulted.

Secondly, the ventilation arrangements are wrong — the exhausts from the engine room come up under the awnings and are hence deflected down into the living spaces."

Engine room and boiler room temperatures averaged 128° for 13 days in AMAZON. AMBUSCADE was similar. The ventilation was soon improved and AMBUSCADE undertook trials of the improved ventilation system later.

The Engineer in Chief reported on the 1928 endurance trial:
"The paper confirms previous reports and observations. Thornycroft has a reputation in this department of turning out a sounder job than Yarrows when given a free hand."
"AMBUSCADE's endurance is considered to be overstated by some 500 miles."
The results of this steaming trial were generally satisfactory as the new machinery was reasonably reliable and durable. The problems of vibration and poor ventilation in machinery spaces were relatively easy to resolve and with the exception of ACHERON's high pressure machinery, all the A to I's were reliable vessels.

BUILDING DETAILS

Name	Builder	Ordered	Laid Down	Launched	Commissioned
AMAZON	Thornycroft	12.6.24	29.1.25	27.1.26	5.5.27
AMBUSCADE	Yarrow	12.6.24	8.12.24	14.1.26	9.4.27

A pre-war shot of AMAZON. She still carries the light shields for her 4.7" mountings.

AMAZON (D39)

AMAZON commissioned for service with the Atlantic Fleet on 5.5.27 and after making alterations at Portsmouth between 3.6–14.7.27, she undertook lengthy trials and work-up until 12.27. She then briefly served with the Atlantic Fleet before returning to refit between 28.2–10.4.28. She then participated with AMBUSCADE on a constant speed cruise from Portsmouth to Pernambuco, Buenos Aires, Valparaiso, Callao, Panama, Kingston, Bermuda and Fayal before returning to Portsmouth on 15.8.28.

AMAZON finally arrived in the Mediterranean for service with the 3rd Destroyer Flotilla (D.F.) on 8.12.28, after repairs and refit at Portsmouth from 3.9–25.10.28. She was to serve with this Flotilla until 4.31, before entering Reserve as tender to the cruiser FROBISHER on 7.5.31. She then refitted at Portsmouth until 28.5.32.

AMAZON was to spend the next two years in Irish waters, punctuated by a maintenance period at Devonport between 1.1–11.2.33. However, on 9.8.33 at Queenstown, the destroyer CRESCENT, whilst manoeuvreing alongside, damaged the AMAZON. She was repaired at Portsmouth and then commissioned with a reserve crew as a tender to HMS VICTORY until 5.34. After a further refit, AMAZON became a tender to HMS VERNON at Portsmouth until 8.36 when the retubing of her boilers was begun in the Dockyard. The work must have had a very low priority, as she was to be in Dockyard hands for some 31 months and did not re-commission into the Vernon Flotilla until 14.2.39 and remained on these duties until mobilisation on 10.8.39.

AMAZON spent the first eight months of the war in the Channel as a unit of the 18th Flotilla, until she was attached to the Home Fleet for the duration of the Norwegian Campaign. She then reverted to the 18th Flotilla until 8.40, when she rejoined the Home Fleet's 12th D.F. This Flotilla was transferred to Western Approaches Command the same month and became part of the Belfast Anti-Submarine Strike Force later the same month. She escorted convoys and Armed Merchant Cruisers to Iceland. During 11.40 she joined the 3rd Escort Group, still in Western Approaches Command. After a brief refit at Glasgow in early 1941, AMAZON remained on escort duties for the next year until she again refitted, this time at Liverpool between 1-3.42.

During 4.42, she escorted Convoy PQ14 to Kola Inlet, arriving with the cruiser EDINBURGH and eight merchant ships on 19.4.42; 16 of the vessels had turned back because of bad ice conditions and the others had been lost to U-boat attack. On the return journey EDINBURGH was torpedoed by U456 and the destroyers Z24 and Z25 in the Barents Sea and was scuttled two days later. AMAZON, BULLDOG, BEAGLE and BEVERLEY protected her as best they could. AMAZON having been damaged during the three day action landed her wounded at Seidisfjord in Iceland, before returning to the Clyde for damage repairs.

Repairs completed, she was allocated to the Greenock Special Escort Force, escorting a convoy to St John's, Newfoundland the next month, before participating in the 'PEDESTAL' convoy to Malta during 8.42. In an action where only four of the 16 merchant ships reached Malta, AMAZON came through unscathed. 9.42 saw AMAZON providing part of the escort for the cruiser force covering the passage of PQ18 to Russia. During 10-11.42, she participated in the convoy operations supporting Operation 'TORCH' and participated in the assault on Oran.

AMAZON then refitted at Troon until 4.43 before operating in West African waters between 6–10.43, when she rejoined Western Approaches Command. However, during repairs at Rosyth during 10.43, AMAZON's material condition was discovered to be so poor that she was relegated to the status of a target ship for aircraft. She continued on these duties throughout 1944. She then acted as a vehicle for damage control tests by the SORF, Forth area until relegated to Reserve during 6.45. She arrived at Troon on 6.10.48 for scrapping.

AMBUSCADE as trials vessel for the Squid A/S mortar in Loch Long during 7.43. (National Maritime Museum N31308)

AMBUSCADE (D38)

On commissioning, AMBUSCADE joined the Atlantic Fleet for four months from 5.27. She then had the first of many periods in Dockyard hands for alterations and modifications at Chatham until 14.12.27. She then rejoined the Atlantic Fleet and participated in the fleet's cruise between 4.1-29.2.28.

After a brief refit at Chatham, she joined AMAZON on the special cruise to South America between 10.4-16.8.28. AMBUSCADE refitted at Chatham between 27.8-13.10.28. Three days later, she re-commissioned as a unit of the 3rd D.F. of the Mediterranean Fleet, where she served until 8.29. In that month, she suffered damage from practice torpedoes to both propellers and her starboard shaft. She repaired at Malta between 21.8-24.10.29.

During 1.30, she relieved VANSITTART in the 4th D.F. of the Mediterranean Fleet for the next eight months. This was followed by eight months of turbine repairs at Malta between 14.8.30-4.3.31. She then returned to the U.K. and acted as tender to the destroyer leader MALCOLM at Sheerness, as part of the Nore Reserve between 1.5.31-10.31. A further refit followed at Chatham between 21.10.31 and 31.5.32 and on 22.6.32 she recommissioned for service in Irish waters until 8.33.

AMBUSCADE then refitted between 5.10-9.12.33 to act as tender to H.M.S. VERNON. She undertook these duties until replaced by ACHERON in 2.37. By this time her turbines must have been in poor condition as the refit at Portsmouth which started on 22.2.37 was suspended on 18.10.37 as her turbines needed replacement. She was then laid up in this state until the outbreak of war, when the work was resumed. AMBUSCADE did not finally re-commission until 27.5.40, some 39 months after the refit started.

After a brief work-up at Portland, although allocated to the 16th D.F. at Harwich, she participated in the evacuation of the 51st Highland Division from St. Valery, near Le Havre. On 10.6.40, AMBUSCADE was hit by shore-fire, which required 17 days repairs at Portsmouth. She then joined her Flotilla and undertook anti-invasion duties along the east coast. Whilst assisting the trawler TURQUOISE, which had been damaged by bombing, she was herself machine gunned by aircraft.

On 3.9.40, after escorting convoy OA205, AMBUSCADE transferred to the 12th D.F. at Greenock for service with the Home Fleet. However, turbine problems then recurred and she repaired on the Tyne between 13.9-8.11.40. When she rejoined her Flotilla, it had been re-allocated to Western Approaches Command. After a period of duty based on Iceland, she refitted on the Clyde between 9.5-7.8.41.

She did not remain on station long, as condenser trouble caused a further period of repair at Portsmouth between 15.10.41-10.1.42. On her return to escort duties, she escorted convoy PQ14 to North Russia and returned to Iceland with QP9. However, further repairs at Liverpool until 18.6.42 more or less finished her front-line career, as she then became an aircraft target ship for the next nine months, punctuated by a refit between 30.11.42-23.1.43.

On 4.6.43 she left Portsmouth for the Clyde on completion of her modifications to fit the newly developed Squid anti-submarine mortar. She then undertook trial and training duties on the Clyde for the next year. A refit was completed between 16.6-8.44 and she then continued her training duties for the next two months. On 15.10.44, AMBUSCADE left Milford Haven for the Mediterranean to undertake deep sea trials, returning to the Clyde on 9.11.44.

AMBUSCADE then remained on the Clyde undertaking training duties until relieved in the Training Squadron during 6.45. She finally entered Reserve at Barrow on 27.8.45. A career blighted by engineering problems was ended when AMBUSCADE was made available for disposal on 23.11.46. She was then handed over to BISCO for scrapping and arrived at the yard of the West of Scotland Shipbreaking Co. Ltd., Troon during 3.47 for demolition, after a period of service as a shock-trials vessel.

THE 1927 PROGRAMME

On 22.7.26, whilst the AMAZON and AMBUSCADE were completing and before they had started their trials, the Assistant Chief of the Naval Staff requested reports on AMAZON and AMBUSCADE and stated that the Controller wanted a staff requirement by 11.26.

The proposed staff requirement for the new design for a destroyer and leader were discussed at a conference held on 11.8.26.

The parameters of the new design were to be:
(i) Endurance was to be the highest priority as it directly affected displacement. Endurance was to be
  a) 1,500 miles at 16 knots
     plus
  b) 12 hours' steaming at two-thirds power. (The two prototypes had allowed for (a) plus 8 hours steaming at two-thirds power.)
(ii) Main Armament:
  The primary armament was to be the torpedoes (21" MK IV's) in two quadruple Mk 1 mounts. The remainder of the armament was to consist of four 4.7" guns (40° of elevation except No. 2 gun with 60° of elevation*). Two pom-poms and Lewis guns completed the vessels' armament.
  *This weapon, then under development, was never fitted.
(iii) Auxiliary Equipment:
  (i) To be fitted with two-speed destroyer sweep (T.S.D.S.) for minesweeping.
  (ii) ASDIC to be fitted (this was later rescinded).
  (iii) The depth charge complement to consist of 4 chutes, 2 throwers and 8 depth charges
(iv) Speed = 33 knots full load.

The vessels were not to be fitted for minelaying, as the new vessels were to replace vessels in the flotillas, where torpedo armament predominated.

The vessels were to be characterised by the following:
(1) The ability to withstand ocean weather and maintain fleet-speed in moderate weather.
(2) They were to be fitted with a large bridge.
(3) Accommodation for their complement was to be superior to existing leaders and the vessels were to be suitable for hot weather operations.

The displacement of the proposed vessels would be approximately 1,325 tons standard and 1,600 tons full load — an increase of 170/200 tons over the 'W' class. This increase was due to the additional fuel to be carried, greater torpedo armament and to strengthening due to the greater gun elevation. (As a rule of thumb 3 tons of displacement can be saved by a reduction of one ton of oil carried.)

The proposed leader would be similar to the destroyer version, except that her armament would be five 4.7" (40° except No. 2 gun 60°), and she would have a larger bridge and greater accommodation due to her larger complement and the need to improve habitability compared with present leaders. T.S.D.S. was not to be fitted. No estimates of displacement were given until the sketch design had been accepted.

The Staff Requirement was issued on 20.12.26, and accepted by the Sea Lords' meeting the next day. It was also agreed that a sketch design be prepared. The First Sea Lord (Beatty) ordered the Controller to prepare the design on 19.1.27.

The Staff Requirement was similar to the earlier Staff Requirement drawn up during August 1926, except:
(1) Oil Fuel to be 425 tons.
(2) To be fitted for fuelling at sea (after recent experiments with RFA PRESTOL and a destroyer).
(3) Speed to be 31¾ knots at full load.
(4) Full load displacement to be around 1,735 tons (less than AMAZON's 1,800 tons, but greater than AMBUSCADE's 1,600 tons).

On 30 March 1927 as a result of the trials of AMAZON and AMBUSCADE it was shown that the Staff Requirement for endurance could be met by 325 tons of oil. Accordingly the design proceeded on the basis of 350 tons of oil carried. (This must be seen in hindsight to have been a mistake, as British destroyers in World War II were notoriously short ranged compared with their U.S.N. counterparts.)

In April 1927 a two boiler design was proposed, but was not proceeded with (see the section on B Class destroyers).

The Sketch design was submitted.

THE NEW DESIGN — THE ACASTAs — COMPARED

Detail	'W'	AMAZON (THORNYCROFT)	AMBUSCADE (YARROW)	NEW DESIGN 2.5.27	NEW DESIGN 31.10.27
Length (PP)	300'	311' 9"	307'	310' 6"	312'
Length (WL)	309'	319'	319'	320'	323'
Breadth	29'6"	31'6"	31'	31'6"	32' 3"
Mean Draft					
(standard)	7'11"	9'2"	8'	8'6"	8'6"
(deep)	9'6"	11'3"	10'	10'3"	10'3"
Displacement					
(standard)	1,112 t	1,352 t	1,173 t	1,330 t	1,330 t
S.H.P.	27,000	41,446	32,795	34,000	34,000
Speed (Trials)	34¼ Knots	37.96 Knots	37.19 Knots	35 Knots	35 Knots
Speed (Deep)	30¾ Knots	33¾ Knots	33¾ Knots	31½ Knots	31½ Knots
Oil Fuel	370 t	433 t	385 t	350 t	350 t
Complement	127	145	145	145	152
Endurance (n.m.) at 15 knots	3,210	3,310	3,400	3,500	3,500
ARMAMENT					
4.7" 30° (rounds per gun)	4 (140)	4 (190)	4 (190)	3 30° (190) 1 60° (290)	3 30° (190) 1 60° (290)
2 pdr pom-pom	2 (100)	2 (100)	2 (100)	2 (100)	2 (100)
21" T.T.	2 x 3	2 x 3	2 x 3	2 x 4	2 x 4
Minesweeping Equipment	—	—	—	T.S.D.S.	T.S.D.S.

As can be seen the ACASTAs were a derivative of Thornycroft's AMAZON. Notice has been taken of the attempt to improve the AA qualities of the vessels, with the provision of a 60° elevation weapon. This was not actually achieved until the 'S' class destroyers of 1942/43.

In October 1927 Goodall (later D.N.C.) as head of the destroyer section had to reduce the standard displacement after the weights had been reassessed.

	Sketch Design Tons (5.27)	Calculated Weight Tons	Rounded Up Legend Tons
Equipment	77	90	90
Armament	123	133	135
Machinery	500	505	505
Hull & Dynamos	630	600	600
	1,330	1,328	1,330

Tenders for the eight destroyers were invited on 27.9.27, followed by that of the Leader a week later. The tenders were received on 8.11.27 and 15.11.27 respectively.

The tender prices for the vessels were as follows:

Tenders for 2 Vessels			First Vessel £	Second Vessel £
Hawthorn Leslie	£439,880	Hull each	92,870	93,520
		Machinery	125,890	125,890
		Equipment	6,903	6,902

Scotts	£440,600	Hull each Machinery Equipment	88,700 131,400 6,689	88,350 121,400 6,689
Swan Hunter	£441,140	(only information surviving)		
John Brown	£441,566	Hull each Machinery Equipment	96,700 123,808 6,778	96,370 123,808 6,778
Armstrong	£441,700	(only information surviving)		
Vickers	£441,924	(only information surviving)		

Tenders for 1 Vessel

Swan Hunter	£224,820	(only information surviving)
Armstrong	£225,400	(Hull £99,700 Machinery £124,500, Other £1,200)

The whole group of destroyers were ordered on 6.3.28 except ACHERON, which was ordered on 29.5.28. ACASTA was inclined on 14.11.29 and her G.M. calculated at 3.05 (light). Her trials four days later found that
(1) She was generally vibration free, except at 23 knots, when there was synchronism between her propellers and the hull.
(2) Engine room ventilation was generally better than AMAZON and AMBUSCADE.
ACHERON was delivered late because of problems with her high pressure machinery.
Admiral H. F. Oliver, C.inC. Atlantic Fleet criticised the design as being "big, too big and costly. A small amount of speed, that can only be achieved in calm seas, should be taken away and sea worthiness encouraged. Sea worthiness is a basic for screening the fleet and convoy protection — speed only predominates when the destroyer is on torpedo duties".

OIL CONSUMPTION

During 1931 certain modification were made to the turbines, reports from the Mediterranean having shown a considerable variation between destroyers — ACASTA and ACHATES were more economical than ACTIVE, ARROW and ANTHONY, their consumption being nearly half a ton per hour at 16 knots. 1.2 tons at 25 knots and one and a half tons at 30 knots.

THE LEADER-CODRINGTON

The vessel differed from her destroyers in the following ways:
(i) Fifth 4.7" mounted (between funnels).
(ii) Bridge increased in size to accommodate Captain 'D' and his staff.
(iii) No T.S.D.S. fitted.
(iv) The general arrangement was as the CAMPBELL class, except where modified by the adoption of the 4.7" Q.F. guns.
No. 2 gun was to have 60° elevation, but subsequently all mountings had 30° elevation only.
Swan Hunter and Wigham Richardson were awarded the contract on 6.3.28, with Parsons turbines and machinery manufactured by Wallsend Slipway under licence.
Sources: Ships Covers 454 and 454A.

1927 PROGRAMME

Name	Builder (Engine Builder)	Ordered	Laid Down	Launched	Completed
CODRINGTON	SWAN HUNTER (Wallsend Slipway)	6. 3.28	20. 6.28	8. 8.29	4. 6.30
ACASTA	JOHN BROWN	6. 3.28	13. 8.28	8. 8.29	11. 2.30
ACHATES	JOHN BROWN	6. 3.28	11. 9.28	4.10.29	27. 3.30
ACHERON	THORNYCROFT (Parsons)	29. 5.28	29.10.28	18. 3.30	13.10.31
ACTIVE	HAWTHORN LESLIE	6. 3.28	10. 7.28	9. 7.29	9. 2.30
ANTELOPE	HAWTHORN LESLIE	6. 3.28	11. 7.28	27. 7.29	20. 3.30
ANTHONY	SCOTTS	6. 3.28	30. 7.28	24. 4.29	14. 2.30
ARDENT	SCOTTS	6. 3.28	30. 7.28	26. 6.29	14. 4.30
ARROW	VICKERS-ARMSTRONGS Barrow	6. 3.28	20. 8.28	22. 8.29	14. 4.30

CODRINGTON at Malta in 1930 as leader of the 3rd Destroyer Flotilla on her first commission.

CODRINGTON (D65)

On completion CODRINGTON worked up in the English Channel before joining the 3rd Destroyer Flotilla on the Mediterranean Station as leader, between 7.30-6.31. She then returned home to have alterations to the turbines made at Devonport, which were completed by 30.6.31. The work involved having her turbines lifted and blading modified.

CODRINGTON then returned to the Mediterranean until 6.37, her service being punctuated by a refit at Devonport between 9-10.32 and a collision with ACASTA off Malta on 12.6.34, which left CODRINGTON little damaged. The final months of service in the Mediterranean included Non-Intervention patrols off the Spanish Coast after the start of that country's Civil War.

A period in reserve at Devonport on her return to the U.K. was followed by a refit. CODRINGTON was then attached to the Royal Navy Engineering College (R.N.E.C.) at Keynsham during 1938/39 and completed her final pre-war refit during 8.39.

On 3.9.39 CODRINGTON was flotilla leader of the 19th D.F. based at Dover, where her first duties were to help escort the B.E.F. to France during 9-10.39. This was followed by normal escort and patrol duties in the English Channel and North Sea. On 4.12.39, CODRINGTON took His Majesty King George VI from Dover to Boulogne and returned with him six days later.

During 2.40, CODRINGTON was allocated as leader of the First D.F. at Harwich, taking up her duties on 6.3.40 on completion of a short refit. A month later CODRINGTON and the rest of her flotilla were loaned to the Home Fleet for duties off Norway. On 28.4.40, whilst on reconnaissance off Narvik Fjord, she wore the flag of Admiral of the Fleet the Earl of Cork and Orrery.

However, on 10.5.40 with the German invasion of the Low Countries, CODRINGTON and her flotilla were recalled to Harwich and evacuated Princess Juliana, Prince Bernhard, their daughters and suite from Ijmuiden three days later. A fortnight later CODRINGTON was embroiled in Operation Dynamo — the Dunkirk evacuation, undertaking eight trips and rescuing some 5,800 troops as well as 33 survivors from the Belgian ship ABOUKIR on 28.5.40. CODRINGTON suffered no damage at Dunkirk and immediately participated in the evacuation of British forces from Le Havre on 12-13.6.40, when 11,000 men were taken off.

Her end came unexpectedly, whilst undergoing a boiler clean at Dover on 27.7.40. She was attacked by German Me109's that approached from the landward side of the harbour. One bomb fell alongside CODRINGTON, which was moored against the depot ship SANDHURST in the submarine basin. The concussion of the water-hammer in this restricted area broke her back. The centre portion of the destroyer collapsed entirely and she quickly settled to the bottom. Luckily only three men were slightly wounded. SANDHURST was also badly damaged. Portions of CODRINGTON's wreck were still to be seen on the beach by the promenade in 1947.

ACASTA is remembered for her loss on 8.6.40 with her sister ARDENT, whilst defending the aircraft-carrier GLORIOUS from the German battle-cruisers SCHARNHORST and GNEISENAU off Norway.

ACASTA (H09)

After commissioning at Clydebank on 14.2.30, ACASTA joined the 3rd D.F. in the Mediterranean Fleet until 5.37. This period of service was interspersed with refits at Devonport between 30.8-29.10.32 and 29.4-3.7.35 and at Gibraltar between 24.11-20.12.33.

However, whilst on exercises off Malta on 12.6.34, ACASTA was in collision with her leader CODRINGTON and required repairs at Malta until 27.7.34. Her last eight months' service in the Mediterranean were largely spent on refugee and anti-intervention duties off the Spanish Mediterranean coast, with the start of the Spanish Civil War.

On her return to the U.K., ACASTA refitted at Devonport between 1.5.37-11.4.38, which included the installation of ASDIC equipment. She then joined the 7th D.F. and served for the remainder of 1938 in Irish waters, before starting a short refit at Devonport between 3.11.38-17.1.39. On completion of this, she became the emergency destroyer at Plymouth until the outbreak of war and assisted Vickers-Armstrongs in the trials between 2.3-13.3.39 of electro-acoustic equipment for the Argentine cruiser LA ARGENTINA, then being built by that company.

ACASTA spent the first eight months of the war as a unit of the 18th D.F. first based at Plymouth for duties in the Channel and then after a short docking period at Devonport between 20.12.39-5.1.40 in the Western Approaches. The highlights of this period were escorting the cruiser AJAX, damaged at the River Plate engagement, into Plymouth on 31.1.40 and when she escorted the newly rebuilt battle-cruiser RENOWN from Devonport to the Clyde.

On 10.4.40 she was ordered to Scapa on attachment to the Home Fleet and spent the period up until her loss on escort duties in Norwegian waters. She escorted the cruiser PENELOPE, badly damaged by hitting a rock, from Norway to the Clyde between 9-16.5.40.

On 8.6.40, ACASTA with her sister ARDENT were the close escorts for the aircraft carrier GLORIOUS, which had been independently routed from Norway to Scapa. The German battle-cruisers SCHARNHORST and GNEISENAU, on a sortie against British shipping, had already sunk the transport ORAMA (19,840 tons), the tanker OIL PIONEER (5,666 tons) and their escort the trawler JUNIPER, when they came upon the squadron. ACASTA was quickly damaged by a shell, which exploded in the forward messdeck, causing damage to the hull and the pom-pom magazine. ACASTA was later immobilised by a direct hit in the engine room and sank by the stern. Her fate and that of GLORIOUS and ARDENT was not known for several days. Only 43 survivors from GLORIOUS and 3 from the destroyers were picked up. Commander Glasford, seven officers and 153 ratings were listed as killed or missing from ACASTA. Leading Seaman Carter was ACASTA's sole survivor. The vessel lies in position 68°45'N 04°30'E.

ACHATES (H12)

ACHATES is remembered for her loss on 31.12.42, whilst protecting Convoy JW51B on passage to North Russia. At 09.20 hours on that date, gunfire was observed in the centre of the convoy and ACHATES altered course to screen the convoy with smoke. 25 minutes later she was hit by a shell that fell short on the port side, holing the forward shellroom, magazine and stokers' messdeck. This caused several casualties and put the Type 271 radar and port Oerlikon out of action.

ACHATES in 7.42 following her rebuilding and conversion to an escort destroyer after being mined off Iceland the year before. She did not survive the year.

ACHATES was hit again at 11.18 hours on the fore end of the bridge, which was wrecked. All the bridge and wheelhouse crew were killed or wounded and 'B' gun and crew put out of action. Two minutes later she was straddled twice and received a third hit which destroyed the seamens' washroom. Her speed was reduced to 12 knots and in order to bring the damaged port side higher out of the water, she sailed on a course parallel to the convoy. ACHATES, however, still screened the convoy with smoke. When the action was broken off, ACHATES was listing 15° to port and was well down by the head. The trawler NORTHERN GEM was requested to take ACHATES in tow, but before this could be accomplished, ACHATES rolled over on her beam ends at approximately 13.00 hours and sank in position 73°03'N 03°42'E. NORTHERN GEM picked up 81 survivors.

ACHATES had been lucky to survive until this date, as some 17 months earlier, at 02.58 hours on 25.7.41, she had struck a British Mk XXSB mine in position 64°25'N 12°40'W, off the south east coast of Iceland.

The explosion resulted in the disappearance of the fore end of the vessel forward of No. 40 Bulkhead including 'A' gun. The vessel was flooded forward of No. 59 Bulkhead, but her machinery and propellers remained undamaged.

Twelve minutes after the explosion ACHATES's engines were re-started and she moved astern at 8 knots to clear the minefield. After two hours the engines stopped due to fuel contamination and the vessel was taken in tow by ANTHONY and both vessels arrived at Seidisfjord at 01.00 hours on 26.7.1941. ACHATES lost 63 missing presumed killed and 25 injured.

After receiving temporary repairs and shoring of bulkheads, ACHATES left Seidisfjord at 20.00 hours on 7.8.41 in tow of the tug ASSURANCE and escorted by ANTELOPE. However, during a storm on 10.8.41, various longitudinals fractured and cracks developed in the upper deck plating. The vessels put into Skaglefjord in the Faroes the next day. After additional shoring was added to the vessel, she left on 21.8.41 for the Tyne, where she arrived at 23.00 hours on 24.8.41. Repairs took over 8 months. On recommissioning, she served as an unattached vessel at Greenock, until 5.7.42, when she joined the Greenock Special Escort Division until her loss. She mainly operated in the Arctic, but during 10/11.42 went south to participate in the 'TORCH' operations.

ACHATES had been completed during 3.30 and then served with the 3rd D.F. in the Mediterranean until paid off at Devonport during 3.37 (on 4.4.32 she suffered a minor collision with ACTIVE off St Tropez). ACHATES was attached to the 1st A/S Flotilla in 11.37 until 6.38. After a further period in reserve, ACHATES was attached to the 6th Submarine Flotilla at Portland until fleet mobilisation in 8.39. She then operated with the 18th D.F. of the Portsmouth Command from the outbreak of war and later the 16th D.F. at Harwich, until her transfer to the 4th Escort Group during 11.40. She then operated with this Escort Group until her mining some 8 months later.

ACHERON was blighted by problems with her high pressure boilers and was soon relegated to second line duties. (W.S.P.L.)

ACHERON (H45)

ACHERON was fitted with experimental high pressure (500 lb/in^2) Thornycroft boilers and was plagued with mechanical problems for the whole of her career. She commissioned on 12.10.31 for special trials with the Atlantic Fleet until 14.6.32 when she was taken in hand for the first in a series of modifications. She re-commissioned on 19.10.32 having been transferred to the 3rd D.F. on 29.8.32. Meantime the 3rd D.F. served in the Mediterranean until early 1935 but ACHERON suffered with mechanical problems throughout the whole commission. In October 1935, she re-commisnioned with the 3rd D.F. for further Mediterranean service but on 3.12.35 she exchanged crews with the WESSEX. She returned to the UK early, arriving at Portsmouth on 17.6.36 to refit and undertake special full power trials "In order that her future service can be decided on, and in view of the necessity for obtaining early information in connection with new construction."

After commissioning on 24.6.36 for special full power trials she refitted at Portsmouth between 17.7.36 and 8.2.37. On that date she commissioned with a special complement as a replacement for AMBUSCADE in the Vernon Flotilla, undertaking local duties until 10.37. During 6-7.37, ASDIC was fitted at Rosyth.

On 1.11.37 she was damaged in a collision with a barge in Portsmouth Harbour, receiving a 20ft tear on her starboard side above the waterline. She repaired at Portsmouth until 6.12.37. She then acted as the Portsmouth emergency destroyer until 3.38 when a refit and further turbine repairs kept her at Portsmouth until 1.12.38. A month later she relieved WINDSOR as a gunnery training destroyer in the Vernon Flotilla, being under repair at Portsmouth when war was declared.

She then undertook local duties around Portsmouth and in the Channel as part of the 18th D.F. until 12.39. Between 12.12.39 and 23.3.40 further machinery repairs were undertaken at Portsmouth and she was then attached to the Home Fleet's 16th D.F. A high angle 3" gun was fitted on 21.6.40.

She suffered nine near misses from dive bombing 10 miles south of St Catherine's Point on 20.7.40. Repairs were started at Portsmouth on 6.8.40. However, at 16.20 hours on 24.8.40, whilst lying inside BULLDOG against the north-west wall of Portsmouth dockyard, she was hit by a bomb. It penetrated the upper deck above the propeller and then detonated below the deck. Her side plating from frame 151 to the stern, port and starboard, was forced outwards and destroyed from the upper deck to the knuckle abaft of frame 157. Her steering engines were severely damaged; the shield of "Y" gun was wrecked and the front plate buckled. Casualties were two killed and three injured.

The damage repairs were completed on 2.12.40 and ACHERON was then to resume her duties as a gunnery training destroyer. Between 10.12.40 and 16.12.40 her "Y" gun was replaced with a gun from BOADICEA.

The next day, 17.12.40, she was on the trials course between St. Catherine's Point lighthouse and Ventnor Pier, when she was mined. Her bows were blown away and her stern tilted in the air and she sank quickly. Only 19 survivors were picked up from her crew and the 25 dockyard workers aboard. 6 officers and 161 ratings were killed. Thus ended the 9 year career of a vessel that had been plagued by mechanical problems from first commissioning.

ACTIVE (H14)

ACTIVE commissioned on 10.4.30 as a unit of the 3rd D.F. After working-up she arrived at Malta on 18.6.30 for service with the Mediterranean Fleet for the next nine months and then returned to Portsmouth to refit and complete alterations between 18.3.31-30.6.31, before rejoining the Mediterranean Fleet. On 4.4.32 she was in collision with ACHATES (q.v.) off St. Tropez and returned to Malta where repairs took 18 days.

She was to serve in the Mediterranean Fleet for the next five years, punctuated by refits at Portsmouth between 1.9.32-16.12.32 and 1.5.34-29.6.34. ACTIVE undertook patrols off Palestine in 6.36 following communal rioting and served off the Spanish Coast from 9.36 to 1.37.

ACTIVE following her collision with the destroyer WORCESTER off the Kurd Bank on 16.2.37.
(Imperial War Museum HU50556)

On 16.2.37, ACTIVE was in collision for a second time, on this occasion with the destroyer WORCESTER off Kurd Bank after her steering gear had failed when travelling at 24 knots. She received extensive damage and her collision repairs and refit at Malta were not completed until 1.6.37. She then commissioned into the 2nd Destroyer Flotilla as a replacement for HUNTER, which had been mined. (q.v.) She was to serve with this flotilla for the next 16 months, punctuated by a refit at Malta between 29.11.37-12.1.38.

On 8.10.38 she reduced to reserve at Malta, where she remained for the next six months, until she commissioned on 22.4.39 as tender to H.M.S. CORMORANT at Gibraltar. On the outbreak of the war she was part of the 13th D.F. at Gibraltar and served on patrol and escort duties there until the formation of Force 'H' in 6.40. (She refitted at Gibraltar between 24.10.39-1.12.39.) She subsequently participated in the action at Oran, when the French battleship BRETAGNE blew up and the battle-cruiser STRASBOURG was damaged.

ACTIVE returned to the U.K. during 8.40 and joined the 12th D.F. until 11.40, firstly with the Home Fleet and later Western Approaches Command. After refitting at Liverpool between 11.40 and 3.41, ACTIVE rejoined the Home Fleet's 3rd Flotilla and participated in the Denmark Straits patrol. She participated in the BISMARCK sortie in 5.41 and four months later ACTIVE and ELECTRA carried R.A.F. personnel to North Russia.

After briefly refitting during 11.41-12.41, ACTIVE saw further service with 38th Destroyer Division of Force 'H' until detached to escort the invasion force for Madagascar during 4.42. She also served as an escort in flying-off operations of aircraft to Malta during this time.

On 8.5.42, during the Madagascar operations, ACTIVE sank the Vichy French submarine MONGE that had attempted to torpedo the aircraft carrier INDOMITABLE, off Diego Suarez. This was to be the first of four submarines that ACTIVE was to help to sink during the next 18 months.

At the end of the Madagascar operations during 9.42, ACTIVE returned to South African waters where she served for the next two months. She sank U179 by gunfire and depth charges west of Cape Town on 8.10.42.

She then returned to the U.K. and commenced a refit, which was not completed until 4.43. ACTIVE was on passage to join the 13th D.F. at Gibraltar as a convoy escort when, with the frigate NESS, she sank the Italian submarine LEONARDO DA VINCI off Cape Finisterre on 23.5.43. There were no survivors. Less than six months later, on 1.11.43, ACTIVE with the destroyer WITHERINGTON, the sloop FLEETWOOD and an aircraft sank U340 east of Ceuta, rescuing all but one of the U-boat crew. During 10.44, ACTIVE returned to the U.K., where she served briefly with the 1st D.F., at Portsmouth, before refitting between 11-12.44. ACTIVE then returned to the 3rd D.F. at Alexandria during 1.45 until the end of the war. She participated in mopping-up operations against German garrisons holding out in the Greek Islands. She captured a landing craft on 28.2.45 and participated in the assault on Piscopi.

After service with the Gibraltar Local Defence Flotilla until 9.45, ACTIVE returned to the U.K. and reduced to reserve at Barrow during 12.45. Whilst laid-up, she was used in ship target trials, until made available for breaking-up 7.7.47. She was scrapped the next month by the West of Scotland Shipbreaking Company Ltd. at Troon. Thus passed one of the few inter-war destroyers to serve the whole of the war without appreciable damage.

ANTELOPE (H36)

On 10.4.30, ANTELOPE commissioned as a vessel of the 3rd D.F., serving with the flotilla in the Mediterranean until 5.37. She was however under modification at Malta between 27.2.31-1.6.31 and her crew took over the destroyer WOLSEY whilst the repairs continued. She served off the Spanish coast between 9-10.36 and 11.36-11.2.37.

On this date, she collided with her sister ship ACTIVE and the destroyer WORCESTER, damaging her fuel tanks. Repairs at Malta were completed on 27.3.37. A further refit was undertaken at Portsmouth

ANTELOPE dressed overall on 19.5.37 shortly before joining the Portsmouth Local Flotilla at the end of her fleet career.

between 5-6.37, following which she joined the Portsmouth local flotilla attached to H.M.S. VERNON until 9.38. She was in full commission during the Munich crisis (24.9-11.10.38), when under the orders of the C. in C. Nore she operated an observation patrol in the Straits of Dover to check on the return of the German armoured ship DEUTSCHLAND to Germany. She then returned to her duties with the Portsmouth local flotilla until 9.8.39 when she was attached to the Reserve Fleet for the inspection by H.M. the King.

On the outbreak of war, she joined the 18th D.F. of the Channel Force and undertook patrol and escort duties until 4.40. On 5.2.40 she sank U41 in the South West Approaches. She was subsequently attached to the Home Fleet during the Norwegian campaign, and escorted the French cruiser EMILE BERTIN to Scapa, after the latter had been damaged by air attack off Namsos on 19.4.40. However, on 13.6.40 she collided with the destroyer ELECTRA off Trondheim and was under repair on the Tyne until 19.8.40, when she joined the 16th D.F. at Harwich.

She was allocated as a convoy escort for operation "MENACE" — the expedition to Dakar. However, she did not reach Dakar as the cruiser FIJI was torpedoed on 1.9.40 and ANTELOPE escorted her back to the Clyde.

ANTELOPE was then based at Greenock, first with the 12th D.F. between 9-11.40 and then joined the 4th Escort Group to 3.41. On 2.11.40, when off north-west Ireland she sank U31, capturing 5 officers and 38 ratings.

During 3.41, ANTELOPE was allocated to the 3rd D.F. of the Home Fleet. On 24.5.41 she escorted the battle-cruiser HOOD and battleship PRINCE OF WALES during their action with the BISMARCK and following the HOOD's loss she searched for her pitifully few survivors. She returned to harbour on 29.5.41, after escorting the aircraft carrier VICTORIOUS. In 8.41 she participated in the Spitzbergen operation.

After a brief refit between 12.41-1.42 ANTELOPE joined the 23rd Escort Group and operated in the South-West Approaches for the next four months. On 1.4.42 she arrived at Gibraltar after escorting convoy WS17 and was allocated to the 18th D.F. of the South Atlantic Command, but was retained in the Gibraltar area until 8.42.

She participated in flying-off operations to Malta with U.S.S. WASP on 20.4.42 and again with WASP and EAGLE on 9.5.42 and on 15.7.42. She also participated in operations "HARPOON" and "PEDESTAL" — escorting convoys to Malta in 6.42 and 8.42. During operation "HARPOON" she towed the torpedoed cruiser LIVERPOOL, escorted by the destroyer WESTCOTT, to Gibraltar.

ANTELOPE finally joined the South Atlantic Station on 30.8.42 with her arrival at Bathurst, Gambia. She then escorted troop convoys for two months, until she was allocated as an escort of KMF1A — one of the assault convoys for the North African landings on 8.11.42. She continued to escort "TORCH" convoys, joining the 13th D.F. of the Mediterranean Fleet in 1.43.

During 1943/44, ANTELOPE was employed on escort duties at Gibraltar, participating in the invasion of Sicily on 10.7.43 and the unsuccessful hunt for U617, which had sunk the destroyer PUCKERIDGE 40 miles off Europa Point, Gibraltar on 6.9.43.

She returned to the U.K. during 8.44 and entered Category 'C' Reserve on the Tyne on 3.10.44. This could have been because of her physical condition or due to the manpower crisis then being experienced. She remained in reserve on the Tyne, until handed over to BISCO and was towed to the yard of Hughes Bolckow Ltd at Blyth for demolition, where she arrived 28.1.46.

ANTHONY (H40)

After commissioning on 11.3.30 for service with the 3rd D.F. of the Mediteranean Fleet, ANTHONY remained with this Flotilla for seven years, until she transferred to local duties at Sheerness during 3.37. Beside the usual fleet exercises, visits and refits, ANTHONY spent between 7.36-3.37 off the Mediterranean coast of Spain. On 13.12.36 she investigated a Soviet steamer on fire in position 36°42'N 0°12'E.

After six months on local duties at Sheerness, ANTHONY refitted at Chatham between 28.9.37-2.3.38 and emerged as a gunnery firing ship and the emergency destroyer at the Nore. However, on 16.1.39, ANTHONY collided with the auxiliary sailing barge LEONARD PIPER. Refit and repairs at Chatham were not completed until 18.2.39. One of her final duties whilst at the Nore was to act as observation vessel in the Straits of Dover (see ANTELOPE).

On 15.3.39, ANTHONY recommissioned as part of the Vernon Flotilla, but on 4.8.39 she joined the 18th D.F. at Portland on the fleet's Mobilisation. ANTHONY spent the first eight months of the war on convoy protection duties at first in the Channel, and then the east coast, before returning to the Channel. These duties were with the 18th Flotilla to 10.39, briefly the 23rd Flotilla and then the 16th D.F. at Portsmouth.

After repairs at Portsmouth between 15.4-20.5.40, ANTHONY was immediately embroiled in the Dunkirk evacuation, carrying over 3,000 soldiers to the U.K., before being damaged by a near-miss on 30.5.40. Repairs at Portsmouth were completed on 18.6.40. ANTHONY rejoined the 16th D.F. that had been transferred to Harwich on anti-invasion duties, during 7.40 before being loaned to the Home Fleet briefly. She then transferred to the 12th D.F. at Greenock for three months until 11.40. ANTHONY joined the 4th Escort Group on 18.11.40 until 13.2.41, when she entered refit at Glasgow after being damaged by splinters during an air raid there.

On completion of her refit, ANTHONY joined the 3rd D.F. of the Home Fleet and after covering some minelaying operations in the Denmark Straits and the Faroes, she escorted the HOOD and the PRINCE OF WALES seeking the German battleship BISMARCK. Detached for fuel the day before the action in which HOOD was sunk on 24.5.41, ANTHONY was recalled and escorted the damaged PRINCE OF WALES to Hvalfjord, Iceland.

ANTHONY spent the next five months on escort duties. The principal events were to escort her sister ACHATES into Seidisfjord, after the latter had strayed into a British minefield off Iceland on 25.7.41, her participation in the destruction of coal installations and radio station on Spitzbergen the following month and escorting convoys PQ1 and QP1 during 9.41.

After refitting on the Humber between 23.10-31.12.41, ANTHONY was allocated to the 38th Destroyer Division of Force 'H', arriving at Gibraltar during 1.42 after escorting convoy WS15. She escorted convoy WS16 during 2.42 and participated in two flying-off operations to Malta during 3.42. ANTHONY was detached to escort the convoys for operation 'IRONCLAD' — the landings at Diego Suarez on 5.5.42. Late on the evening of that day, ANTHONY transported 50 marines to land in the harbour to act as a successful diversion.

ANTHONY, as part of the Eastern Fleet's 2nd D.F., spent the next four months on local escort duties between South Africa, Madagascar and East Africa, before sailing for Philadelphia with DUNCAN on 17.9.42 as escort for the torpedoed battleship ROYAL SOVEREIGN. After being diverted to New York, they arrived on 22.10.42 to participate in escorting the American convoy UGF2 for the 'TORCH' landings. DUNCAN and ANTHONY after refuelling at Ponta Delgada, then joined convoy CF7, before escorting the cruiser SHROPSHIRE to the Clyde, where they arrived on 15.11.42. ANTHONY then refitted on the Thames between 21.11.42-4.3.43.

ANTHONY with limited war modifications. She is serving with the 2nd D.F. in the Indian Ocean between 5.42-9.42.

After working up, ANTHONY joined the 13th D.F. at Gibraltar and was based there until 9.44. She participated in operation 'HUSKY' on 10.7.43 and with the destroyer WISHART sank U761 on 24.2.44 by gunfire and depth charges, after the submarine had been damaged by aircraft off Tangier.

On her return to the U.K. she operated as an Aircraft Target Ship in the Channel for the next six months, before undertaking her conversion for that duty between 13.3-6.6.45. She then operated out of Lamlash and later Douglas, Isle of Man on these duties.

ANTHONY was replaced by RAPID as an air target on the Forth during 1.46 and entered Category 'C' Reserve on 27.3.46. In the same month she was approved for scrap, but six months later she was allocated to the Ship Target Trials Committee for trials. She was handed over to BISCO on 21.2.48 before finally arriving at the yard of the West of Scotland Shipbreaking Co Ltd., at Troon during 5.48 for demolition.

ARDENT (H41)

After commissioning at Chatham on 23.4.30, ARDENT worked-up and after exchanging No 4 mounting, which was defective, she sailed to join the Mediterranean Fleet's 3rd D.F. on 19.5.30. ARDENT was to serve three commissions in the Mediterranean, the first from 6.30 to 12.30, which was terminated by defects. ARDENT arrived at Malta on 31.10.30 and was taken in hand on 1.12.30 for repair. She did not recommission from Reserve until 4.11.31. She then completed two further commissions on the Mediterranean station between 11.31-1.34 and 2.34-4.37.

On her final commission, ARDENT spent much of her time in Spanish waters — between 8.9-17.10.36 and 29.11.36-4.37 off the Spanish Mediterranean coast. This final period of service was as SNO Barcelona.

ARDENT is seen displaying the submarine contact flag during a pre-war exercise.

Refitted at Sheerness between 14.4.37 and 20.4.38, which included the fitting of ASDIC, ARDENT spent the next 18 months as the Devonport emergency destroyer. She commissioned with a full complement during the Munich crisis (24.9-11.10.38) before refitting at Devonport between 17.10-15.11.38. A brief cruise to Torbay and Dartmouth as a Boys Training Vessel was followed by another period in dockyard hands until 23.8.39. She was then brought forward to war complement and on the outbreak of war was a unit of 18th D.F. at Portland as part of the Channel Force. The Flotilla was re-allocated to the Western Approaches six days later and ARDENT operated on convoy escort duties until 4.40. With her sister ACASTA, and the destroyer WHITSHED, she was one of the escorts for the cruiser AJAX on her return to Plymouth on 31.1.40.

On 12.4.40, ARDENT arrived at Scapa as escort for a convoy from the Clyde to Norway and was retained in the naval forces supporting the Norwegian operations. She was under repair from 6-19.5.40 for the replacement of her A/S dome and between 22-25.5.40, she escorted the troop transport ULSTER PRINCE transporting forces to occupy the Faroes, returning to Greenock on 29.5.40.

On 8.6.40, ARDENT was posted missing after joining the carrier GLORIOUS as close escort, with her sister ACASTA. It transpired that on the afternoon of 8.6.40, the group were intercepted by the German battle-cruisers SCHARNHORST and GNEISENAU and despite the efforts of the destroyers to lay smoke and fire torpedoes, all three vessels were sunk with a heavy loss of life. There were only two survivors from the ARDENT, who were picked up by a German seaplane and made prisoners of war. Ten officers and 142 ratings were lost. The wreck lies in position 68°45'N 04°30'E.

ARROW (H42)

ARROW commissioned at Chatham on 23.4.30 for service with the 3rd D.F. of the Mediterranean Fleet. She served with this fleet until 5.3.31, when she, ACTIVE and CODRINGTON returned to the U.K. to have their turbines lifted and the blading modified. This was successfully undertaken at Chatham between 18.3.31 and 11.7.31, when she sailed for Gibraltar and service with the 3rd Flotilla until 4.37. She refitted at Chatham between 30.8-17.10.32, at Malta between 23.10.33-9.1.34 and finally at Sheerness between 29.4-29.6.35. On her final Mediterranean commission she operated off southern Spain between 8-10.36, 11-12.36 and 1-4.37.

After a refit at Sheerness between 6.5-17.7.37, which included the fitting of ASDIC, ARROW entered Reserve at Sheerness, as a tender to the cruiser CARDIFF until 3.38. On 2.3.38, she commissioned as part of the Portsmouth Local Flotilla until 8.39, refitting at Portsmouth between 11.11.38-10.1.39. During the Munich crisis (24.9-11.10.38) she was in full commission and shadowed the DEUTSCHLAND on her passage through the Channel.

After attendance at the Reserve Fleet Review, during 8.39, ARROW was a member of the 18th D.F. until 11.39 — undertaking local escort duties. However, a recurrence of turbine defects entailed repairs at Devonport between 24.10.39-10.1.40. She then joined the 16th D.F. and immediately on recommissioning provided the escort with ACHATES and ANTHONY for the battleship ROYAL SOVEREIGN from Portsmouth to Halifax, where the battleship was to undertake trade protection duties. However, ARROW needed further repairs between 31.1-9.3.40, which further restricted her activities.

ARROW was then attached to the 12th D.F. of the Home Fleet at Scapa on 22.4.40. She was not to remain on station long, as four days later she encountered a trawler wearing the Dutch flag off Andalsnes. On closing, the trawler (which was, in fact, the German SCHLESWIG (SCHIFF 37), shortly afterwards sunk by the cruiser BIRMINGHAM) opened fire and rammed ARROW. ARROW was holed above the waterline and was repaired at Middlesbrough between 29.4-13.5.40. On returning to service, ARROW spent the next month in Norwegian waters, rescuing 80 survivors from a group of Norwegian vessels bound for Thorshavn and sunk by bombing.

A short refit and change of armament was undertaken at Sheerness between 26.6.-3.7.40, before a brief return to the 16th D.F. — now based at Harwich. On 24.7.40 she was re-allocated to the Western Approaches Command and based at Greenock. ARROW's next period of service was marked by numerous rescue operations including rescuing two survivors from a Greek vessel on 27.8.40, 10 survivors from the Greek POSEIDON on 13.9.40 and rescuing survivors from the EMPIRE WIND on 13.11.40. The next day she was ordered to assist the SAN DEMETRIO lying off Achill Head and successfully escorted her into the Clyde two days later. (SAN DEMETRIO's crew had re-boarded the vessel, after she had been abandoned following damage by the German cruiser ADMIRAL HIPPER when she attacked convoy HX84 on 5.11.40).

ARROW's poor mechanical condition forced her to spend between 17.11.40-14.1.41 under repair at Barclay Curle's yard on the Clyde. Repairs to all of ARROW's boilers and other repairs were undertaken. She then spent between 1-5.41 escorting Icelandic convoys to Scapa, Aberdeen and the Clyde. However, her boiler problems recurred and further repairs were undertaken at Chatham between 2.5-21.6.41, which included a complete refit and the retubing of her superheaters. At this time she was a member of the 3rd D.F. When 107° 4½ miles off Flamborough Head at 20.07 hours on 21.6.41, whilst on passage north, ARROW exploded a mine 50ft off the starboard quarter about station 160. The vessel was severely shaken and all lights failed due to the failure of the dynamos. Hull damage was slight, mainly leaking rivets and local dishing of plating with some bulkheads distorted. However, much damage was caused to the supports of the auxiliary machinery. Steam was maintained on No. 1 boiler and ARROW steamed at 9 knots to Middlesbrough, where she arrived the next day. 'Y' gun and torpedo tubes were out of action. Repairs at Smiths Dock, Middlesbrough were not completed until 20.11.41.

ARROW, although officially a unit of the Western Approaches Command, was to see world wide service during the next year. She arrived at Alexandria on 29.12.41 and immediately undertook escort duties. On 12.1.42 she was narrowly missed by a U-boat torpedo in position 31°40′ N 27°46′ E. ARROW and HERO unsuccessfully hunted the submarine.

ARROW was then sent as a reinforcement for the Eastern Fleet's 2nd D.F. and arrived at Addu Attoll on 4.4.42. She operated with the fleet until the need of yet another refit forced her to make passage to Durban, where she repaired between 21.5-2.7.42. She then operated in the Indian Ocean, off the coast of South Africa and at Diego Suarez until 9.42. During this month she transferred to the West African station for a further two months until further boiler problems necessitated her return to the U.K. and an extensive refit at Middlesbrough between 18.11.42-26.3.43.

When working-up at Scapa on 10.4.43 in heavy weather, she got entangled in one of the booms there and made passage to London, where she was repaired by Green & Silley Weir at Blackwall between 13.4-30.5.43. She returned to Scapa to work-up between 3-14.6.43 and then left for the Mediterranean and service vessels the 13th D.F. at Gibraltar.

ARROW manoeuvring in a dock basin during 5.43. She was to survive barely two months, before being extensively damaged in an explosion at Algiers on 4.8.43. *(Imperial War Museum, FL3570)*

At 15.40 hours on 4.8.43, whilst ARROW was lying at Algiers, the FORT LAMONTEE, which was on fire, blew up and set ARROW on fire causing heavy casualties. ARROW was towed to Gibraltar to undertake temporary repairs, arriving on 18.9.43. However, she left for Taranto on 19.11.43, arriving there under tow eight days later. ARROW was reduced to Category 'C' Reserve on 8.11.43.

Progress on ARROW was very slow and her poor material condition, the availability of more modern destroyers and the chronic manning crisis being experienced by the Royal Navy at this time must have been factors in the decision of 17.10.44 not to continue the repairs and for her hull to be stripped. Her hulk lay at Taranto until broken up there in 5.49.

SAGUENAY AND SKEENA (R.C.N.)

The Canadians had been debating for a considerable time during the late 1920s over what vessels were to replace the R class destroyers PATRIOT and PATRICIAN then in Canadian service. Despite a popular campaign to have the vessels constructed in a Canadian shipyard, the decision was made to order the vessels in the U.K. On 27.2.28, two Canadian Officers — Engineer Commander Philips and Commander Beard examined the plans of the ACASTA class and made the following alterations which were considered desirable to fit the vessels for Canadian conditions:
(i)   The forecastle and upper deck were to have the sheer strake extended above the deck plating. Scuppers were to be fitted.
(ii) An additional accommodation ladder to be fitted forward.

(iii) Two additional whalers were to be fitted abaft the bridge and a motor-boat and a whaler with dinghy below, further aft.
(iv) As they were to be fitted with an oil fired boiler for heating accommodation spaces, coal heating stoves were therefore not required.
(v) Bathrooms to be fitted for crew.
(vi) Bridge to be streamlined, without affecting general construction of the bridge.

On 4.6.28 tenders were invited by the Canadian High Commission for the construction of the two vessels. Thornycrofts replied in a tender dated 31.7.28. However, it was to be several months before the order was placed and much further discussion on the design took place meantime.

On 24.9.28, when tenders were being considered, Thornycrofts put forward the design for a special destroyer, substantially the same as the Admiralty 'A' class, which had
(i) Heavier scantlings and more robust construction.
(ii) Maximum speed of 35 knots.
(iii) A radius of action of 5,000 miles at 12 knots.
(iv) Loaded displacements of between 1,600/1,700 tons.
(v) High tensile steel and galvanising used during construction.

However in a letter dated 15.1.29, S. V. Goodall, Deputy Director of Naval Construction, gave further particulars of the Canadian vessels:
(i) Generally to follow the 'A' class.
(ii) To be 4ft shorter due to reduced machinery space.
(iii) To have 32,000 S.H.P. and not to be superheated.
(iv) The reduction in the weight of machinery to be matched by heavier scantlings.
(v) Oil fuel capacity to be increased for the Esquimalt-Panama run.

Goodhall doubted that, with less power and heavier scantlings, they would reach the speed of the 'A' class. This proved to be a correct assessment.

The vessels were finally ordered on 29.2.29 from Thornycrofts.

	Tender Price for two vessels	Contract Price for two Vessels
Hull	£186,600	£186,680
Main Machinery	£271,600	£275,860
Auxiliary Machinery	£14,002	£14,002
Twin Motors for Steering Gear	—	£640
Cost of Trials	£5,600	£2,800
Supply of Spares	£24,828	£12,414
	£502,630	£492,396

The vessels were named SAGUENAY and SKEENA on 27.6.29 and the armament agreed at:
    4 x 4.7" Mk IX guns
    2 x 2 pdr pom-pom (later replaced by multi barrelled 0.5 machine gun)
    4 x Lewis guns

During construction some small changes were made to the vessels, with the overall length being increased by $1\frac{3}{4}''$ to 321' $1\frac{3}{4}''$, the beam by $1\frac{1}{8}''$ to 32' 8" and the depth of the vessels by $\frac{1}{8}''$ to 19' $\frac{11}{16}''$.

BUILDING DETAILS

NAME	LAID DOWN	LAUNCHED	COMMISSIONED
SAGUENAY	27.9.29	11.7.30	22.5.31
SKEENA	14.10.29	10.10.30	10.6.31

Source: Cover 465

SAGUENAY (R.C.N.) (D79)

SAGUENAY was the first ship ordered for the Royal Canadian Navy and commissioned on 22.5.31. After a short work-up she made passage to Halifax with her sister SKEENA. SAGUENAY was to be based on the Canadian east coast, with occasional detachments to the Caribbean until the outbreak of the war. After briefly escorting local convoys from Halifax, she was assigned to the America and West Indies Station at the end of 9.39. She was based at Kingston, Jamaica and undertook patrol duties for the next three months. On 23.10.39, she intercepted the German tanker EMMY FRIEDERICH in the Yucatan Channel. The German vessel scuttled herself.

SAGUENAY under construction at the Thornycroft yard on 6.2.30. She is still 5 months away from launch.
(National Maritime Museum G7152)

In 12.39, she resumed her local escort duties at Halifax until 16.10.40 when she sailed for the U.K. to join E.G.10. She was to serve with the group barely six weeks, as on 1.12.40, whilst escorting convoy HG47, she was torpedoed by the Italian submarine ARGO some 300 miles west of Ireland. The torpedo destroyed SAGUENAY's bow and killed 21 of her crew.

She made passage under her own power to Barrow in Furness, where a new bow was fitted and completed on 22.5.41. Leaving Greenock the next day, she crossed the Atlantic and immediately joined the Newfoundland Escort Force where she remained for the next eighteen months. She escorted the battleship PRINCE OF WALES to Placenta Bay for the meeting of Mr. Churchill and President Roosevelt between 9-12.8.41.

SAGUENAY dressed overall at the Coronation Review of 1937. *(National Maritime Museum N3005)*

In 1.42, whilst escorting convoy ON52, SAGUENAY received such severe storm damage that she barely reached St. Johns, N.F., and required three months in dockyard hands to repair the damage.

On 15.11.42, she was in collision with the Panamanian freighter AZARA, south of Cape Race, Newfoundland and her stern was destroyed by the explosion of her own depth charges. There were no fatalities. SAGUENAY was towed in and docked at Saint John, New Brunswick, where her stern was sealed. No permanent repairs were undertaken and SAGUENAY was used as a static training vessel at H.M.C.S. CORNWALLIS from 10.43 until paid off on 30.7.45. She was broken up in 1946 by International Iron and Metal Co., Hamilton, Ontario.

SKEENA (R.C.N.) (D59)

After commissioning at Portsmouth on 10.6.31, SKEENA worked up with her sister SAGUENAY and they sailed together for Halifax thirteen days later. The sisters arrived at Halifax on 29.6.31. On 3.7.31 SKEENA set out with the destroyer VANCOUVER for Esquimalt, where the vessels were to form the Western Division of the Royal Canadian Navy.

SKEENA underway off the west coast of Canada in 1933. She was based at Esquimalt between 1931 and 1938.
(National Maritime Museum N3077)

The highlight of this period of service, which did not end until 4.38, was on 22.1.32 when en route to the Panama Canal, SKEENA was directed to Acajutla in San Salvador, where she hosted some British subjects during civil disturbances in that town. SKEENA and SAGUENAY also represented Canada at the Coronation Naval Review held at Spithead on 20.5.37.

On her replacement in the Western Division by FRASER in 4.38, SKEENA returned to Halifax and operated from that port until the outbreak of war. A pleasant duty was to convey the King and Queen to Pictou in 7.39.

Immediately on the outbreak of war she conveyed the C. in C. America and West Indies and his staff to Bermuda, returning to Halifax on 11.9.39, the day after Canada declared war. She then joined the Halifax Escort Force and was employed as a local escort for HX and HXF convoys until 5.40.

She arrived at Plymouth on 31.5.40 and joined the Western Approaches Command. However, she spent the next month escorting convoys of troops and evacuees from ports on the west coast of France. On 14.10.40, she rescued 230 survivors from the torpedoed A.M.C. CHESHIRE, north-west of Ireland and put a steaming crew on board, which kept the vessel moving until salvage tugs took over. SKEENA continued convoy escort duties until 3.3.41, when she left to refit at Halifax.

On completion of her refit in 6.41, she joined the newly formed Newfoundland Escort Force and operated with this group and later the Mid Ocean Escort Force until 12.42, when she started a much needed refit at Halifax. The highlight of this period of service was the sinking of U588 on 31.7.42, with the corvette WETASKIWIN, whilst protecting convoy ON115.

The refit was completed in 4.43 and she immediately joined Escort Group C3 for the next 10 months on North Atlantic convoy duties. After a further refit between 2-4.44 and after crossing the Atlantic with convoy SC157, she joined other 'River' class destroyers of the Canadian Navy as part of Escort Group 12 at Londonderry. The Group was given the task of protecting Channel traffic after the invasion. During 7.44, she participated in Operation 'DREDGER' — the attack on the surface escort vessels for U-boats from Brest. After an action in Audierne Bay, SKEENA collided with the QU'APPELLE and suffered 14 wounded and needed several weeks in dockyard hands.

During 9.44, she joined Escort Group 11, which was detailed to destroy U-boats on passage to and from Norway. On the night of 24/25.10.44, she took refuge in Reykjavik harbour from a gale. At about 02.00 hours the next day, she dragged her anchors and was driven ashore off Viday Island. 15 of her crew died when the carley float they were using to get ashore capsized and was lost.

SKEENA was declared a constructive total loss and was sold during 6.45. She was then refloated and broken-up locally.

THE 1928 PROGRAMME

On 1.8.28 Admiral Madden, the First Sea Lord, in a memo to the Controller outlined the requirements for the new Fleet destroyer. The new vessels were to be repeat ACASTAs, but armed with quadruple torpedo tubes, 4.7" guns with 30° elevation and to be stiffened for the future fitting of High Angle guns. Other features of the vessels were to be a maximum speed in deep loaded condition of 31.5 knots; fire control arrangements similar to AMAZON and AMBUSCADE; stowage of 10 weeks' stores and the fitting of two depth charge throwers and a rail. The Twin Speed Destroyer Sweep, however, was not to be fitted.

Technical discussions were limited to whether the fourth 4.7" mounting should be suppressed and a third set of quadruple torpedo tubes fitted. The D.N.C. replied on 10.8.28 to the effect that such a design would save some 20 tons in displacement and speed would be increased by $\frac{1}{4}$ of a knot. The saving of £12,600 for the gun mounting would be more than offset by the extra cost of £14,000 of the quadruple torpedo tubes. The D.N.C. added, however, that if the torpedo tubes were fitted, it would be difficult to operate the sweep gear because of space restrictions and it would make the passage of spare torpedoes impossible. The crew of the torpedo tubes would also be affected by the blast of No. 3 gun. The D.N.C. wrote to the Controller at this time (8.28) concerning the proposed vessels:

"The increase in size of the ACASTA class has been due to heavier armament, more ammunition, increase in complement with a higher standard of comfort and accommodation than formerly"; in addition the vessels "have a more elaborate system of Fire Control".

The D.N.C. outlined the weight increases of 185.83 tons of the new design over the repeat W's, which consisted of a new type of gun (8.6 tons), more complex depth charge throwers (1.47 tons), greater ammunition stowage (31.50 tons), the stowage of minesweeping gear not carried in the W's (18.4 tons) 10 tons more fuel oil, greater machinery weight (88 tons) to give more power, heavier searchlights (0.36 tons), dynamos (6.00 tons), and greater bridge structure (1.5 tons).

Originally it had been intended to fit a new mark of torpedo tubes, but as the new tubes were insufficiently tested by the tender date of 23.10.28 the same type of tubes as the ACASTAs were fitted.

The design was approved under Board Minute 2517 of 18.10.28.

There was no separate design for the leader. The additional accommodation for the extra 20 crew carried by the leader was provided by dropping a 4.7" gun and building extra superstructure in lieu. This had been agreed on 25.8.28 in order that the leader and destroyers had the same tactical radius. The details of the vessels tendered for were:

Dimensions: Length (OA) 323', Breadth $32\frac{1}{4}$', Draft $8\frac{1}{2}$'

Displacement was 1,330 tons. The 34,000 S.H.P. rated machinery drove the vessels at 31.5 knots in full load condition.

Weights: (tons)	Destroyer	Flotilla Leader
General Equipment	100	106
Armament	122	104
Machinery	505	505
Hull	603	615
Standard Displacement	1330	1330

There was no Board Margin and oil fuel stowage was 380 tons.

Armament:
- 4   4.7" QF (30°) (3 only in the leader) with 190 rounds per gun
- 2   2 pounder pom-pom with 500 r.p.g.
- 4   Lewis guns with 2000 r.p.g.
- 8   Torpedo tubes
- 15 Depth Charges with 2 throwers and 1 rail

The orders for the 9 vessels were placed on economic and unemployment criteria at the following yards at the following contract prices:

John Brown (Clydebank)	: 2 Destroyers:	£213,460 for each vessel
Palmers (Jarrow)	: 2 Destroyers:	£214,000 for each vessel
Swan Hunter (Wallsend)	: 2 Destroyers:	£214,500 for each vessel
Hawthorn Leslie (Hebburn)	: 2 Destroyers:	£218,000 for each vessel
Vickers-Armstrongs Ltd (Barrow)	: The Leader:	£219,800

During construction minor modifications were made to BLANCHE when it was agreed during 10.29 that this vessel be fitted to carry an officer of Captain's rank and the cabin flat was re-arranged to accommodate 2 extra petty officers and 2 extra stewards and to BULLDOG which was fitted with one C XIII (60% elevation) mounting and a new type of ASDIC dome.

PROPOSALS BY YARROW AND THORNYCROFT FOR A TWO BOILER DESIGN

Both companies when tendering for the vessels offered a two-boiler design for the leader instead of the three-boiler design. Few details of the Thornycroft design survive.

The Yarrow design was characterised by the fact that the two-boiler design would have the same surface area as the Admiralty design, the vessel was to have the same dimensions as that proposed, with the minimum of design changes in the area of the boiler/engine rooms.

Yarrow stated that the two boiler design would save 14ft in length and 20 tons in weight; the profile of the vessel would be reduced, as the two-boiler design only required one funnel instead of two; pipe work would be simplified and the safety of the vessel improved due to the reduced floodable length of the engine room and boiler room.

The Admiralty criticised the designs in the following ways:
(i) The additional 4.7" gun was badly positioned amidships.
(ii) In the Thornycroft vessel, the accommodation was below standard.
(iii) Their stability would not be as good as the Admiralty design.

Beside the last point the criticisms are not serious and did not affect the basic 2 boiler design.

The D.N.C. made further comments:

"The leader was attractive from a design point of view, but the matter would need careful consideration in all the aspects of staff and machinery considerations and it is not proposed to consider such a leader for the present programme."

The idea did not die, however, as £10,000 was authorised under the 1929/30 Estimates for the construction of an experimental boiler. The Admiralty wished to wait until the trials of the experimental boiler had been completed before deciding whether the reduction in weight in a two boiler ship should be devoted to building smaller vessels of similar fighting powers to the present type, or vessels of the same size as the present type, but of greater fighting power.

Development was slow and it was not until the 1936 Estimates and the 'J' class that the two boiler design was adopted. British destroyers were to prove mechanically reliable during the war, unlike their German counterparts, which suffered repeated mechanical breakdowns.

BUILDING PARTICULARS

NAME	BUILDER (Engine Builder)	ORDERED	LAID DOWN	LAUNCHED	COMPLETED
KEITH	VICKERS-ARMSTRONGS	22.3.29	1.10.29	10. 7.30	20.3.31
BASILISK	JOHN BROWN	4.3.29	19. 8.29	6. 8.30	4.3.31
BEAGLE	" "	4.3.29	11.10.29	26. 9.30	9.4.31
BLANCHE	HAWTHORN LESLIE	4.3.29	29. 7.29	29. 5.30	14.2.31
BOADICEA	" "	4.3.29	11. 7.29	23. 9.30	7.4.31
BOREAS	PALMERS	22.3.29	22. 7.29	11. 6.30	20.2.31
BRAZEN	"	22.3.29	22. 7.29	25. 7.30	8.4.31
BRILLIANT	SWAN HUNTER &	22.3.29	8. 7.29	9.10.30	21.2.31
BULLDOG	WIGHAM RICHARDSON (Wallsend Slipway & Engineering Co.)	22.3.29	10. 8.29	6.12.30	8.4.31

KEITH (D06)

On 9.6.31 KEITH commissioned at Chatham as leader of the 4th D.F. and was then stationed with the flotilla in the Mediterranean between 6.31-8.36. She refitted at Chatham 4.9-18.10.33.

However, on 24.8.36 whilst on passage between Gibraltar and Portsmouth to refit, she was in collision in thick fog in the Channel with the Greek steamer ANTONIS G. LEMOS. She arrived at Portsmouth the next day and her refit and collision repairs were completed there during 12.36. However, modifications to the ventilation arrangements to the petrol compartment meant that she did not recommission until 13.2.37. She then spent six months in Reserve at Sheerness.

On 14.8.37 she temporarily commissioned as the leader of the 6th D.F., whilst their leader FAULKNOR was under repair following collision damage. KEITH spent between 8-9.37 off the Biscay ports and the next month she undertook patrols from Gibraltar. She returned to Sheerness on 4.11.37 and commissioned for service in the Reserve Fleet at the Nore fifteen days later. She remained in Reserve until 9.5.38, when she started a refit at Chatham, which was completed on 16.6.38. She then commissioned for temporary service with the 4th D.F. in place of BEAGLE and served in Home waters until 1.39.

A fine pre-war view of KEITH whilst acting as leader. (Leaders did not carry pennant numbers.)

On 17.1.39 she paid off and re-commissioned with the crew of ELECTRA for temporary service with the 5th D.F. until 4.39. During this period she undertook patrols from Gibraltar covering the end of the Spanish Civil War. She refitted at Chatham between 11.5-15.7.39, re-commissioning into the Reserve Fleet on 31.7.39.

On 3.9.39 KEITH was serving with the 17th D.F. with the Home Fleet, but almost immediately joined the Western Approaches Command and undertook anti-submarine patrols based at Milford Haven until 29.10.39 when she made passage to Harwich.

On 3.11.39 KEITH became leader of the 22nd D.F. with the Polish Destroyer Division, BOADICEA, GRIFFIN, GREYHOUND and GIPSY. However, she soon entered Devonport Dockyard for propeller repairs, which were not completed until 10.1.40. During 2.40, KEITH relieved CODRINGTON as leader of the 19th D.F. and served with this Flotilla until her loss.

On 5.3.40 she escorted BOADICEA whilst the latter was towing the damaged tanker CHARLES F. MEYER into Southampton. She then undertook patrol and escort duties until the German invasion of the Low Countries on 10.5.40, when she was immediately embroiled in evacuation duties. On that day, with BOREAS, she escorted the cruisers ARETHUSA and GALATEA and two merchant ships carrying gold bullion from Ijmuiden to the U.K. She then returned to the Hook of Holland to evacuate British and Dutch troops.

KEITH was subsequently involved in the Dunkirk evacuation, assisting the destroyer WHITLEY, which was sinking off Nieuport after being bombed on 25.5.40. Five days later KEITH sailed from Dover to Dunkirk, where she arrived at 20.00 hours, returning to Dover five hours later with 1,200 troops on board. She then returned to the beaches of La Panne and Braye, passing orders and marshalling small boats. Admiral Ramsey, Flag Officer Dover and Field Marshal Lord Gort, the commander of the B.E.F., conferred on board KEITH during the day of 31.5.40.

Lord Gort and his staff left at 03.00 hours on 1.6.40 and almost immediately the vessel was damaged by machine gun fire. By the time she was damaged by near misses at 07.30 hours, her anti-aircraft ammunition was almost expended and her steering gear damaged. Further air attacks followed and the vessel was straddled; No. 2 boiler room was set on fire and its crew killed by a bomb going down the funnel. She was listing, with her upper decks touching the water and equipment was jettisoned. As no power was available, she was anchored and Admiral Ramsey and his staff were transferred to an M.T.B. for passage to Dover. KEITH was then abandoned and sank at 09.45 hours on 1.6.40. 3 officers and 33 ratings were killed but her Commanding Officer, 7 other officers and 123 men reached home safely by various means. KEITH lies in position 51° 04' 46" N 02° 26' 47"E.

BASILISK spent 8 years with the 4th D.F. of the Home Fleet before being superseded by a "Tribal" in 3.39. (W.S.P.L.)

BASILISK (H11)

Completed on 4.3.31, BASILISK was a unit of the 4th D.F. for the next eight years, in the Mediterranean until 9.36 and then with the Home Fleet until 3.39. At this time she became the emergency destroyer at Devonport before being allocated to the 19th D.F. at Dover on the outbreak of war. BASILISK spent the next eight months on the usual destroyer tasks of escort or patrol, before being detailed on 17.4.40 for duties with the Home Fleet off Norway.

She escorted the battleship RESOLUTION with the destroyers WREN and HESPERUS to Narvik on 24.4.40. Eleven days later she escorted the troopship EMPRESS OF AUSTRALIA to Tjelsundetfjord and gave support for the landing at Bjerkvik on 12-13.5.40 during the operations to capture Narvik.

On 30.5.40 BASILISK was ordered south to participate in Operation 'DYNAMO' — the evacuation of Dunkirk, which already had been proceeding for four days. BASILISK evacuated some 1,115 troops the next day and returned to the beaches of La Panne on 1.6.40 to recover more troops.

However, at 08.15 hours that day, she was dive bombed by 9 aircraft and received a bomb hit at the after end of the boiler-room as well as several near misses. She was severely damaged and immobilised with all her boiler-room and engine room staff becoming casualties. Some four hours later BASILISK was again attacked by several aircraft and six bombs exploded near her.

She sank at 12.13 hours in 24 feet of water off the beaches. Her survivors — 8 officers and 123 men — were picked up by the Belgian fishing vessel LA JOLIE MASCOTTE, which with the destroyer WHITEHALL had earlier tried unsuccessfully to tow BASILISK. She sank in 51° 08' 16"N 02° 35' 06"E.

BEAGLE (H30)

BEAGLE returned to Devonport on 27.8.36 following over five years service with the 4th D.F. in the Mediterranean since first commissioning on 15.5.31. Before returning to the U.K., she had been despatched, at the request of the High Commissioner to Palestine, to Jaffa to aid the civil power during communal unrest.

BEAGLE refitted at Devonport until 16.1.37, when she returned to the 4th D.F., now part of the Home Fleet, until 4.38. Another leisurely refit at Devonport followed between 4.4-17.9.38, when BEAGLE commissioned to replace STRONGHOLD as the attendant destroyer to the aircraft carrier FURIOUS and Capt. (D) of the 15th D.F. of the Home Fleet. She was to be on these duties barely two months before undergoing a further period in dockyard hands between 24.11.38-3.1.39, prior to becoming plane-guard for the training carrier ARGUS for the next four months.

Between 12.4-3.5.39 BEAGLE was under repair at Devonport for minor collision damage with BASILISK before undertaking another two months' spell of plane-guard duties with FURIOUS at Rosyth. A period for docking and repairs at Devonport completed her peace-time service before she joined the 19th D.F. at Dover during 9.39. She served with the flotilla on routine duties until 4.40, refitting at Falmouth between 18.12.39-22.1.40.

BEAGLE was occupied on convoy duties between the Orkneys and Narvik between 4-6.40 and then participated in the evacuation of British nationals from Bordeaux to Plymouth.

On 3.7.40 she was ordered to join the 1st D.F. at Dover and for the next 16 days she undertook patrols by day and night. However, on 19.7.40 she was damaged outside Dover Harbour by JU 87's and sustained damage to her gyro compass and boiler room that required repairs at Devonport until 16.8.40.

On completion, she was retained at Devonport as a unit of the 22nd D.F. serving in the Channel for the next two months. On 14.10.40 BEAGLE was attached to the Home Fleet for escort duties and four days later escorted the carrier ARGUS conveying naval aircraft to Iceland and immediately afterwards escorted a West African convoy. In 2.41, she was transferred to the Western Approaches Command as part of the 4th Escort Group on the Clyde-Iceland convoy run.

During severe weather on 24.10.41 she suffered a broken foremast and other damage which was repaired at Greenock and at the same time she was fitted with radar. However, she was to suffer more extensive weather damage in 12.41 and proceeded to the Tyne for repairs. During this refit she was converted into a short range escort with an early Hedgehog installation and with torpedo tubes modified to fire a one ton depth charge.

She sailed as escort for Convoy PQ14 in 4.42 and while returning with QP11 from Murmansk on 1.5.42, BEAGLE, BULLDOG, AMAZON and BEVERLEY beat off five separate attacks by three large German destroyers armed with a total of ten 5.9"guns and five 5" guns compared with the six 4.7" and three 4" guns of the escorts. BEAGLE suffered minor splinter damage.

Between 5-10.42, BEAGLE returned to the Greenock Escort Force and escorted troop convoys from the Clyde to south of Iceland, where the ocean escort took over. During 10-11.42 BEAGLE was attached to Force 'H' for escort duties for Operation 'TORCH'. She then escorted convoys JW51A, RA51 and JW52 to and from North Russia.

Refitted with improved radar, A.S.W. equipment and improved insulation for Arctic duties, BEAGLE in true navy fashion was of course detached to Freetown, where she operated as a local escort until she returned to the Home Fleet in 11.43. She then made five round trips to North Russia, with the 8th Escort Group between 11.43 and 5.44. Whilst escorting convoy JW58 she participated in the sinking of U355 with aircraft

BEAGLE off the Normandy Beaches carries many war modifications, and the two pounder bow-chaser must have been added for her service there.

from the escort carrier TRACKER on 1.4.44. BEAGLE then participated in Operation 'NEPTUNE' until 19.7.44 escorting landing craft and ships between the south coast of England and the assault area. She destroyed two JU88's when attacked on 22-23.6.44. She then repaired defects at Sheerness between 19.7-9.44 and rejoined the 8th Escort Group on coastal escort duties. Further defect repairs between 12.44 and 2.45 sidelined BEAGLE before she rejoined the 8th Escort Group for a few more weeks. On 11.3.45 she was re-allocated to the Plymouth Command for escort duties in the western English Channel. From 12.4-1.5.45, BEAGLE operated with the Biscay Blockade and attended with BULLDOG the surrender of the German garrison in the Channel Islands on 9.5.45. 15 days later, BEAGLE was taken in hand at Devonport for destoring and reduction to Category 'C' Reserve. She was approved for scrapping on 22.12.45 and a month later on 15.1.46 was handed over to BISCO at Rosyth.

She moved to the Metal Industries shipbreaking yard at the north-west corner of the dockyard two days later.

BLANCHE (H47)

On 9.6.31 BLANCHE commissioned at Portsmouth for service with the 4th D.F. of the Mediterranean Fleet where she was to serve until 8.36. Between 22.4-25.4.35 she carried the flag of the Rear-Admiral Destroyers at Malta.

After refitting, BLANCHE and the remainder of the Flotilla joined the Home Fleet until 3.38. She saw service in Spanish waters between 2-4.37 and 1-3.38, when based at Gibraltar for service in southern Spanish waters.

BLANCHE had the unfortunate distinction of being the first British destroyer to be lost in World War II, when she was mined in the Thames estuary on 13.11.39.

On 6.3.38 in position 37°26′ N 00°05′ E, she was bombed, without damage, by five Nationalist aircraft. She had been closing on the British s.s. SHAKESPEAR to investigate another bombing incident that had just taken place.

After a refit at Portsmouth between 1.4.38-11.6.38, BLANCHE joined the A/S flotilla at Portland until 3.39. However, during the Munich crisis BLANCHE was one of four destroyers that escorted the liner AQUITANIA and the battleship REVENGE in the Channel on 30.9.38. BLANCHE again refitted at Sheerness between 1.4.39-15.7.39 before becoming the emergency destroyer at the Nore.

On the outbreak of war, she joined the 19th D.F. at Dover and for the next two months undertook patrol and escort duties in the Channel and North Sea.

On 13.11.39, BLANCHE and BASILISK were escorting the minelayer ADVENTURE when the latter was mined off the Tongue light vessel in the Thames Estuary. BLANCHE also detonated a mine under the after portion of the ship. The upperdeck was split forward of the after superstructure and all power was lost. The engine room, tiller flat and spirit room were all making water and BLANCHE was taken in tow for Sheerness by the tug FABIA. However, at 09.50 hours the same day she capsized in a position 1 mile 80° NE of the Spit Buoy. A rating was killed and twelve others were wounded. (The mines had been laid, that same night, by the German destroyers KARL GALSTER, WILHELM HEIDKAMP, HERMAN KUNNE and HANS LUDEMAN.) BLANCHE was the first destroyer lost to enemy action during the war.

BOADICEA (H65)

On commissioning on 2.6.31, BOADICEA joined the 4th D.F. and served in the Mediterranean between 7.31-8.36. She was detached to aid the civil power at Haifa and Famagusta between 11.35-1.36 and again at Haifa during 6.36. Her final months in the Mediterranean were spent at Cartagena and Valencia, evacuating British and other nationals from the disturbances that signified the start of the Spanish Civil War. Earlier, on 15.3.35, BOADICEA had been damaged when practising refuelling at sea with the battleship REVENGE; repairs at Gibraltar were completed between 15.3-18.4.35.

After a refit at Portsmouth between 21.8-26.9.36. BOADICEA rejoined the 4th D.F., now part of the Home Fleet, until 3.38. She spent much of this commission in Spanish waters — between 2-3.37, 9-11.37, 1-3.38 at Gibraltar and southern Spanish ports, punctuated by a spell off the Biscay ports between 6-8.37. After a further refit at Portsmouth between 7.4.38 and 11.6.38 BOADICEA rejoined the 4th D.F. of the Home Fleet until 1.39, when the B's were superseded by the newly commissioned ''Tribals''. The next two months were spent on plane-guard duties with the aircraft carriers of the Mediterranean Fleet and then BOADICEA acted as the emergency destroyer at the Nore, until being attached to the Reserve Fleet at Portland in 8.39 for the Review.

On 29.8.39 BOADICEA was stationed at Dover with the 19th D.F. After covering the movement of the B.E.F. to France during 9-10.39, BOADICEA joined the 22nd D.F. for two months at Harwich for the defence of the East Coast before rejoining the 19th D.F. On 4.3.40 she towed the tanker CHARLES F. MEYER into Southampton Water after the latter had been mined. She missed the Dunkirk operations as she was under refit at Chatham from 2.5.40. She was, however, operational by 9.6.40, when she went to Le Havre to help the evacuation of the 51st Highland Division.

She was very lucky to survive the excellent bombing of a group of nine JU87's off the coast between Fecamp and Dieppe at 17.37 hours on 10.6.40. The first attack missed 30 yards to starboard, but the second attack obtained three hits, the first of which penetrated BOADICEA's deck on the starboard side, and entered the engine room and burst. The second bomb exploded in the engine room, killing all the engine room personnel but one. The third bomb entered the after boiler room and passed through the ship's bottom without exploding. The next series of bombs missed to port. The ship stopped and listed to port and orders were given to abandon ship. However, by 17.50 hours it was reported that the fore-end of the after boiler room and the after end of engine room were watertight and being shored up and the list reduced. It was now thought that the vessel could be saved. All depth charges, torpedoes and much of the equipment were jettisoned and a collision mat put over the hole in the boiler room bottom plating. The work was completed at 18.30 hours, but the vessel was shrouded in fog, and it was not until 20.00 hours that AMBUSCADE was sighted and took BOADICEA in tow. The tow was taken over by the tug KROOMAN at 16.20 hours on 11.6.40 and she was towed into Dover at 22.00 hours.

Repairs at Portsmouth to BOADICEA were not completed until 14.2.41, which included fitting of Type 286 radar. She then joined the Home Fleet at Scapa and immediately escorted the fleet in the search for the German battle-cruisers SCHARNHORST and GNEISENAU which had broken out into the Atlantic. On her return from these duties, BOADICEA joined the 4th Escort Group at Greenock for escort duties until 2.42, when the group was dispersed. She then operated as an unallocated vessel in Western Approaches Command until 7.42. During 4-5.42, she escorted convoys PQ15 and QP12 to and from Kola Inlet under severe air and submarine attack. She then joined the Special Escort Division at Greenock during 7.42. After completing the escort of the troopship DOMINION MONARCH from Halifax to Liverpool, BOADICEA refitted on the Clyde between 8.42 and 10.42.

A battered looking BOADICEA, which had a distinguished career before succumbing, with heavy loss of life, to aerial torpedoes off Portland Bill on 13.6.44.

She then escorted convoy KX2 to Gibraltar as part of the 'TORCH' operation. At 16.43 hours on 8.11.42, BOADICEA engaged a French destroyer of the L'ALCYON class off Oran, and was struck by a 5.1" shell which burst in the forward shell room. A hole in the ship's side was immediately plugged and the shell room pumped out.

Three days later she was in company with the troopship VICEROY OF INDIA when the latter was torpedoed and sunk. The BOADICEA rescued 425 survivors and carried them to Gibraltar. On 19.11.42 she arrived at Greenock as part of the escort for convoy MKF1.

Between 12.42-3.43, BOADICEA operated as part of the 20th Escort Group and was part of the escort for the convoys JW51A, JW53 and RA53 to and from North Russia. On 10.3.43, whilst escorting RA53 she was extensively damaged forward by ice; a week later she started repairs on the Clyde, which were not completed until 5.43.

BOADICEA then joined the West African Command at Freetown, escorting convoys between Freetown, Takoradi, Lagos and Capetown. On 19.7.43 she rescued 220 survivors from the m.v. INCOMATI. She left Freetown for the U.K. on 15.9.43 and after a brief period with 8th Escort Group, refitted on the Tyne between 11.43-1.44 being fully converted into an escort destroyer (see Appendix II).

After working-up, BOADICEA rejoined the 8th Escort Group where she served until her loss. She escorted convoys JW57 and RA58 and RA59 during 2-4.44, when she was based at Skaalefjord in Iceland. BOADICEA then made passage to Plymouth and participated in the initial assault convoys for the Normandy Landings. On 13.6.44, whilst escorting a follow-up convoy off Portland Bill, BOADICEA was hit by two torpedoes launched by JU88's and sank in three minutes with only 12 survivors. 9 officers and 161 ratings were lost.

BOREAS saw world-wide service with the Royal Navy, before being loaned in 2.44 to the Royal Hellenic Navy, with whom she served for another 7 years as SALAMIS. (Courtesy R. Wilson)

BOREAS (H77)

BOREAS spent two commissions with her sisters in the 4th D.F. in the Mediterranean between 7.31 and 8.36. She refitted at Portsmouth between 5.9.33 and 18.10.33 and 21.8-26.9.36. The final months of this service BOREAS spent on patrol and evacuation duties off Southern Spain at the start of the Spanish Civil War during 7.36.

The 4th D.F. then joined the Home Fleet during 10.36 for the next 18 months. However, much of BOREAS's service was spent in Spanish waters between 1-3.37, off the Biscay ports between 6-8.37 and off Southern Spain between 8-11.37 and finally 1-3.38. On 6.3.38 BOREAS and KEMPENFELT picked up survivors from the Spanish Nationalist cruiser BALEARES which had been sunk by Government forces off Cartagena. The survivors were transferred to other Nationalist warships.

BOREAS then refitted at Portsmouth between 1.4-11.6.38, and on completion she acted as escort for the venerable Royal Yacht VICTORIA AND ALBERT during the Royal Tour of Scotland between 26.7-4.8.38. During the Munich crisis, BOREAS escorted the battleship REVENGE, which was protecting the liner AQUITANIA on 29-30.9.38. BOREAS was retained in the 4th D.F. between 16.11.38 and 4.39 in place of ASHANTI, which although commissioned in 12.38 did not complete work-up until 4.39.

BOREAS was briefly employed as a plane guard in place of WREN during the summer of 1939, but on the outbreak of war she became a member of the 19th D.F. of the Nore Command. She spent the first months of the war on escort duties along the East Coast and Channel.

However, on 4.2.40, whilst rendering assistance to the stricken minesweeper SPHINX, bombed by German aircraft in the Moray Firth, BOREAS's stern was damaged at the waterline and she was under repair for the next month. On 29.3.40, BOREAS was attached to the 12th D.F. for the next 6 weeks but on 15.5.40 she was in collision with her sister BRILLIANT and received hull damage above the waterline that required repairs on the Thames until 19.6.40.

BOREAS immediately joined the 1st D.F. at Dover but did not remain long. On 25.7.40, BOREAS and BRILLIANT left Dover to attack 'S' boats in the Channel. At 18.00 hours when 3 miles off Dover she was near missed by bombs to port and starboard abreast the after engine room. BOREAS was temporarily immobilised due to the disablement of the boiler room fans and dynamos and through slow flooding of the engine room. She then proceeded at 17 knots by hand steering. However, at 19.35 hours she received two direct hits on the bridge, causing severe structural damage. She was totally immobilised and taken in tow. One officer and 20 ratings were killed.

BOREAS was under repair in Millwall Docks from 30.7.40 to 23.1.41, receiving superficial damage from bomb splinters on 19.1.41.

During the repairs, she was fitted with the equipment to fire a 10 depth charge pattern. After working up at Scapa during 2.41, BOREAS joined the Western Approaches Command as an unallocated vessel. BOREAS and BRILLIANT were then assigned as replacements for DUNCAN and FOXHOUND with the 18th D.F. of the South Atlantic Command. She left the Clyde as escort for the A.M.C. CORMORIN and rendezvoused with the GLENARTNEY and the Canadian leader ASSINIBOINE, the vessels arriving at Gibraltar on 11.4.41. After attending to defects at Gibraltar for 10 days, BOREAS arrived at Freetown on 28.4.41. She then operated in West African waters on escort duties until 10.8.41, when she arrived at Gibraltar to join convoy HG70.

Subsequently, she rescued five survivors from the British steamer ALVA (sunk 19.8.41), 24 survivors from the Norwegian SPIND (sunk 23.8.41), four from the British tug EMPIRE OAK (sunk 22.8.41) and four from the British steamer ALDERGROVE (sunk 23.8.41); BOREAS returned to Gibraltar with these survivors on 25.8.41.

She subsequently refitted at Middle Docks, South Shields 19.9.41-4.1.42. BOREAS then undertook full power trials whilst on passage to Greenock on 6.1.42 before leaving as escort for a W.S. convoy to Freetown four days later. She arrived at Freetown on 25.1.42 and rejoined the 18th D.F. for the next nine months.

After escorting convoys around the Cape, BOREAS arrived at Alexandria on 11.11.42 and immediately participated in Operation STONEAGE — the convoy from Alexandria which lifted the siege of Malta. She then remained in North African waters until 1.43, escorting another convoy to Alexandria before returning to Gibraltar and a brief period of service with the 13th D.F. She then returned to West African waters between 2-6.43.

BOREAS was one of the reinforcements for the Mediterranean Fleet in 6.43 and participated in operation 'HUSKY' — the Sicily Landings. She then returned to the U.K. for an urgently needed refit by Harland and Wolff at Liverpool between 9.43-2.44. Loaned to the Greeks on 10.2.44, she was commissioned on 25.3.44 as SALAMIS. She then returned to Scapa to work-up, but was damaged, which necessitated repairs at Hull by Amos & Smith between 28.4.44 and 13.6.44. After working-up she made passage to Gibraltar, where she operated until 10.44. She subsequently served with the Greeks in the Eastern Mediterranean and Greek waters until the end of the European War.

Returned to the Royal Navy at Malta on 9.10.51, she was subsequently allocated to BISCO. She arrived at Rosyth in tow of the tug MERCHANTMAN on 15.4.52 for demolition by Metal Industries (Salvage) Ltd, before transferring to Charlestown, where demolition was completed.

BRAZEN was in the headlines in 6.39, when she attended the sunken submarine THETIS in Liverpool Bay. She was to survive barely a year longer. (W.S.P.L.)

BRAZEN (H80)

After completion on 8.4.31 BRAZEN immediately joined the 4th D.F. of the Mediterranean Fleet. She suffered severe corrosion, both fore and aft, on the outer bottom of her hull during the first two years of her service. The affected plates were replaced and improved anti-fouling paint used. Although no definite cause was discovered, it was suggested that the problem was connected with the polluted waters at Jarrow, where she lay whilst fitting out.

Her service in the Mediterranean was punctuated by refits at Devonport between 8-10.33 and at Malta starting during 12.33. BRAZEN, with the remainder of the 4th D.F., was attached to the Home Fleet at the close of 1935 and spent the next three years largely in Scottish waters. The close of her service with the 4th Flotilla coincided with her attendance at the abortive rescue operations on the submarine THETIS which had sunk whilst on trials in Liverpool Bay on 1.6.39.

In 8.39 BRAZEN formed part of the 19th D.F. at Dover and remained on station there until 2.40, when she was attached to the Home Fleet. She had unsuccessfully attacked a U-boat on 7.9.39 and on 13.10.39 she picked up three survivors in the Channel from U40 which had been mined and sunk some hours previously.

On 18.2.40, BRAZEN and ENCOUNTER took over the escort of convoy HN12, after its escort, the destroyer DARING, had been sunk by U23 that same day. A few hours later BRAZEN and DIANA picked up 29 survivors from the Norwegian vessel SANGSTAD, which had been sunk 115 miles east of Kirkwall. BRAZEN and BOREAS then escorted the minelayer TEVIOTBANK on a minelaying mission on 21.2.40. During the Norwegian Campaign BRAZEN formed part of the escort for troop convoy NP1 on passage to Namsos on 13.4.40. Two days later BRAZEN and FEARLESS were detached to search Vaagsfjord for a reported submarine. At 10.48 hours that day, a contact was made and an attack undertaken, which six minutes later resulted in U49 breaking to the surface with 41 of her crew being rescued. Her principal deployments were to act as escort for the battleships RODNEY and VALIANT and the battle-cruiser RENOWN to Scapa, where they arrived on 18.4.40, escort the troop transport GUNVOR MAERSK to Namsos with KIMBERLEY and WOLVERINE, escort convoy FP3 to Narvik arriving on 3.5.40 and finally escort three French transports to Tromso with the Polish Brigade. On 28.4.40 she was bombed without being damaged.

At 03.45 hours on 30.5.40, whilst on passage to Harwich, she hit a submerged wreck off the mouth of the Wash and had to proceed to the Humber for docking to repair damaged plating in the region of the after boiler room. Repairs took some five weeks and then BRAZEN joined the First D.F. at Dover.

She was soon in action, as on 8.7.40 she was attacked by 14 aircraft whilst escorting a coastal convoy off Hythe. BRAZEN was undamaged, but the S.S. CORUNDUM was hit and had to return to Dover. BRAZEN's third brush with enemy aircraft was to be fatal. On 20.7.40 she joined the escort of convoy CW7 under severe air attack between Dover and Folkestone. She was near-missed several times, which broke her back and finally she was hit in the engine room and sank at 2040 hours in position 51°01'05"N, 01°17'15"E with the loss of just one of her crew. Her gunners were reported to have destroyed three Stukas.

BRAZEN is seen with her back broken and sinking between Dover and Folkestone, after being near-missed by German bombs on 20.7.40.

The death throes of BRAZEN are shown minutes later in the late evening of 20.7.40. Note the false bow wave.

BRILLIANT is seen prior to her launch on 9.10.30. (Swan Hunter Ltd)

BRILLIANT (H84)

After working-up, BRILLIANT joined the 4th D.F. at Malta on 3.8.31, where she was to spend the next five years. She acted as guardship at Malaga on the outbreak of the Spanish Civil war. The 4th Flotilla was re-allocated to the Home Fleet in 9.36 and spent the following winter on international patrols in the Bay of Biscay. The B's were replaced in the 4th Flotilla by "Tribals" in 1938/39 and on the outbreak of war BRILLIANT was serving with the 19th D.F., based at Dover.

However, a collision with the Dover breakwater on 12.9.39 meant that BRILLIANT was in dockyard hands for six weeks. She then operated on patrol and escort duties in the Channel until the invasion of the Low Countries on 10.5.40.

Two days later, BRILLIANT conveyed a demolition team to Antwerp and returned to Dover with over a hundred evacuees on board. On 15.5.40, whilst making passage to the Hook of Holland, BRILLIANT collided with her sister BOREAS and was under repair at Sheerness until 17.6.40. She then joined the First Destroyer Flotilla at Dover, on disbandment of the 19th Flotilla, following heavy losses at Dunkirk.

At 17.05 hours on 25.7.40, BRILLIANT was ordered to attack S boats outside Dover Harbour and by 17.47 hours was in action with leading enemy vessels but at 18.00 hours was ordered to withdraw. She was attacked at 18.10 hours by eight JU87's, four miles off South Foreland. She was near-missed several times, but two bombs hit BRILLIANT either side of the quarter-deck and passed through the ship without exploding. There were no casualties and two JU87's were lost. Her steering gear was, however, put out of action and she stopped. She flooded rapidly and settled by the stern. The vessel was lightened aft by the jettisoning of depth charges, protective plating and X and Y guns. She was towed into Dover at 19.50 hours and two days later was towed to Chatham for repairs, which were not completed until mid 10.40.

BRILLIANT then re-joined the Home Fleet until 2.41, escorting the newly completed battleship KING GEORGE V from the Tyne to Rosyth en-route to Scapa.

A brief refit followed at Southampton during 3.41, where one set of torpedo-tubes was replaced by a 3" gun and the number of depth charges was increased to 60, two single 20mm Oerlikons were added and Type 286 surface warning radar fitted.

On 12.5.41, she sailed for Freetown escorting the carrier FURIOUS and the cruiser LONDON to Gibraltar on the way. On 4.6.41 LONDON and BRILLIANT intercepted the German supply ship ESSO HAMBURG between Sierra Leone and Brazil and the next day the EGERLAND in the same area. The interceptions were made because of the breaking of the German Naval Enigma code.

BRILLIANT spent the next 10 months based at Freetown on escort and patrol duties, becoming leader of the 18th D.F. from 8.41. She returned home and started a refit at Chatham in 4.42, when her quarter deck 4.7" gun was landed and two more Oerlikons added. BRILLIANT left for the South Atlantic in 5.42 and escorted a convoy to Durban, before returning to Freetown during 8.42 where she remained until 10.42. She then returned to Gibraltar, where she was attached to Force 'H' for the 'TORCH' operation.

On 8.11.42 BRILLIANT's task was to give a close gun support for landing beach 'Y' at Oran. When the Vichy French sloop LA SURPRISE issued out of Oran to attack the invasion force, BRILLIANT intercepted and sank her. 21 survivors were rescued.

Six days later, BRILLIANT was transferred to the Gibraltar Command serving with the Escort Group 61 until the end of 1.43, when she returned to the U.K. to refit and fully convert to an escort destroyer at Portsmouth. These works, which were not completed until 6.43, including the fitting of a Hedgehog, improved radar, increased stowage for 125 depth charges, the removal of the 3" gun and the re-installation of the second set of torpedo tubes.

A war-time view of BRILLIANT showing her many modifictions including radar and Oerlikons. The torpedo-tubes are, however, retained. (Imperial War Museum A27126)

After working up on the Clyde, BRILLIANT rejoined the 13th D.F. as part of the Gibraltar escort force from 8.43 to 9.44. During this period she participated in several 'swamp' operations, during which U-boats were hunted to exhaustion by the employment of large numbers of aircraft and escorts.

After a further refit at Portsmouth, BRILLIANT joined the First D.F. during 11.44 and operated in the Channel. However, she sustained a seriously damaged bow when she was in collision with the Canadian corvette LINDSAY in fog in the Channel. After emergency repairs at Portsmouth, BRILLIANT transferred to Antwerp for long term repairs, which were not completed until 23.4.45.

She then returned to Portsmouth where she was converted to a submarine target and escort vessel, which was completed on 26.5.45. On 7.6.45 she acted as escort for the cruiser JAMAICA taking H.M. King George VI to Jersey.

Between 13.6-11.45 she operated from Holy Loch, escorting surrendered U-boats from various ports around the U.K. In 11.45, BRILLIANT was approved to reduce to Reserve and she was stripped of her radar and communications equipment, before being laid up in the Holy Loch on 29.4.46 in Category 'C' Reserve. Handed over to the Target Trials Committee on 11.9.46 as a trials ship, she was allocated to BISCO on 21.2.48 and was scrapped at Troon by the West of Scotland Shipbreaking Co. Ltd., from 4.48.

BULLDOG (H91)

BULLDOG's most important contribution to the war effort occurred on 9.5.41, when protecting convoy OB318; she, the destroyer BROADWAY and corvette AUBRETIA drove U110 (Kapitan Leutnant Julius Lemp) to the surface and forced her abandonment. Lemp was killed and the survivors were picked up by AUBRETIA. Commander Baker-Cresswell, C.O. of BULLDOG realised that the survivors thought that U110 had sunk and told them nothing, while a party from BULLDOG stripped the submarine of her Enigma coding machine and the day's settings, which proved an immense cryptoanalytical gain for the Allies. U110 sank in tow some time later.

BULLDOG is seen on her trials in early 1931. (Swan Hunter Ltd.)

Previous to this, after commissioning into the 4th D.F. on 9.6.31, BULLDOG had served with this flotilla in the Mediterranean until 8.36. The highlights of this portion of her career were her attendance at the earthquake-stricken island of Erissos in the Aegean between 28.9-1.10.32 and spending the last month of her Mediterranean service off the Southern Spanish coast. BULLDOG refitted at Gibraltar between 12.5-11.6.32, 2-31.1.35 and had a restricted refit at Malta between 1-20.6.36.

The 4th D.F., including BULLDOG returned to the U.K. during 8.36 and she remained with the Home Fleet until replaced by ESKIMO in 12.38. BULLDOG was under repair at Chatham between 28.8-14.10.36 before refitting between 18.11.36-9.1.37. Although nominally a Home Fleet destroyer, BULLDOG spent long periods in Spanish waters — 1-3.37 off Southern Spain, 6-8.37 off the Biscay ports, 9-11.37 on patrol between Gibraltar and Oran and finally between 1-3.38 based at Gibraltar. BULLDOG then refitted at Sheerness between 31.3-4.6.38. During the Munich crisis BULLDOG escorted the battleship RESOLUTION to Scapa Flow on 26.9.38.

On replacement by ESKIMO, BULLDOG joined the Gibraltar Local Flotilla until 3.39, when she replaced WISHART on plane-guard duties for the carrier GLORIOUS, first in the Mediterranean until the outbreak of war and then in the Red Sea and Indian Ocean until 12.39.

After refitting at Malta between 18.1-22.2.40, BULLDOG and WESTCOTT undertook similar duties for the carrier ARK ROYAL between 2-4.40. On the start of the Norwegian Campaign, BULLDOG was ordered to the U.K., where after de-gaussing and repairs to her feed water heater between 19.4-3.5.40 at Devonport, she began operations immediately. Six days later, she stood by the destroyer KELLY after the latter had been torpedoed by the German S31 in the Skagerrak. BULLDOG towed the seriously damaged vessel to the Tyne for repairs at her builders, Hawthorn Leslie. BULLDOG damaged her stern during the tow and was repaired at Swan Hunters between 13-21.5.40.

BULLDOG participated in the early stage of the Dunkirk evacuation, but was withdrawn for repairs after damaging her propellers on 27.5.40. Repaired at Chatham until 4.6.40, she had been in service with the 1st D.F. barely six days when she received severe bomb damage whilst operating off the French coast. She was attacked by six aircraft which secured 3 hits on BULLDOG putting her steering gear out of operation. The first bomb hit the upper deck at Station 104, passed through the engine room and then through the ship's side. Bomb 2 hit the upper deck at Station 94, entered the top of No. 3 boiler and after passing through the generator tubes came to rest on the water reservoir without exploding. The final bomb first pierced the fore side of the after funnel and then passed through the fore-end of the boiler room. This bomb did not immediately explode, but later went off as a result of the heat of the boiler.

After the bombing BULLDOG turned several circles to port, listing heavily to starboard. Later steam was raised and BULLDOG was able to make 15 knots, after the steering gear was made operable at 18.15 hours. She arrived at Portsmouth at 07.00 hours on 11.6.40. Fortunately, there were no casualties. Repairs at Portsmouth were not completed until 2.9.40, after the vessel had received further splinter damage during an air attack on 24.8.40. She then rejoined the First D.F. On 8 and 10.9.40 BULLDOG, BEAGLE, BERKELEY and ATHERSTONE made sweeps off the French coast attempting to intercept German convoys.

After a refit at Cammell-Lairds between 2.1-18.2.41, BULLDOG worked up and joined the 3rd Escort Group: her days as a fleet escort were at an end, and she spent the next 8 months on Icelandic convoy duty, damaging U94 west of the Faroes, with AMAZON and the sloop ROCHESTER, two days before her successful action with, and capture of, U110.

BULLDOG's sides were strengthened during a refit at Fairfields on the Clyde between 24.10.41-10.2.42 and she then rejoined Western Approaches Command as an unattached vessel. She stood by the destroyer RICHMOND, damaged in collision with the American ship FRANCIS SCOTT KEYS in position 63°37'N 22°05'W on 31.3.42.

The Arctic convoys were by this time (5.42) in full operation and whilst escorting QP11 from Murmansk, BULLDOG with AMAZON, BEVERLEY and the corvette SNOWFLAKE fought off five attacks on the convoy by the German destroyers HERMAN SCHOEMANN, Z24 and Z25. The Germans sank the Russian straggler TSIOLKOVSKY, and BULLDOG received splinter damage and AMAZON a hit to her steering gear. BULLDOG was under repair on the Clyde 2.6-14.8.42. She then formed part of the Greenock Special Escort Division and participated in the 'TORCH' operations.

After further repair at Greenock 23.11-14.12.42, she joined the escort for JW51B from Loch Ewe to Murmansk on 20.12.42, but received weather damage and was again under repair on the Clyde 28.12.42-16.1.43. She then escorted Icelandic convoys for the next two months, before repairing at Greenock 29.3-22.4.43 prior to transfer to the West African Command during 5.43. She undertook escort duties between Freetown, Lagos and Gibraltar for the next five months.

BULLDOG off Spithead in 1944 after refitting as an escort destroyer at Portsmouth. Her greatest distinction was the capture of an Enigma coding machine from U110 on 9.5.41. *(Imperial War Museum FL 1816)*

On her return to the U.K., BULLDOG refitted and re-armed as an escort destroyer at Portsmouth 8.11.43-24.5.44. She was working up at Tobermory on 'D' Day, before undertaking local escort duties between the Faroes and the Clyde. On 26.6.44 BULLDOG sank U719 after a long search in the North Channel.

However, on 20.8.44 she was in collision with the frigate LOCH DUNVEGAN in Gourock Bay. BULLDOG was cut through for a distance of 4ft inboard from the upper deck to 3ft below the water line. She was repaired at Ardrossan 17.8-4.9.44.

BULLDOG then operated on local escort duties between the Faroes, Scapa and the Clyde until 11.44, when major machinery repairs were started at Elderslie, Clyde which were not completed until 30.1.45. The last months of the war BULLDOG spent on local escort duties between Plymouth and Irish Sea ports. On 13.5.45 BULLDOG and BEAGLE had the pleasant task of accepting the German surrender of the Channel Islands.

Almost immediately afterwards, BULLDOG entered Reserve at Dartmouth on 27.5.45 and was declared as Category 'B' Reserve on 26.6.45. She left Dartmouth for Rosyth on 27.11.45, arriving two days later. On 13.12.45 she entered Category 'C' Reserve at Rosyth and was approved for scrap nine days later. On 17.1.46 she was delivered to Metal Industries (Salvage) Ltd at Rosyth for demolition, one of the few destroyers to be on active service with the Royal Navy for the whole of the war.

THE 1929 PROGRAMME

In a memo to the DNC dated 7.2.29, the Controller detailed the general features of the proposed leader and destroyers of the 1929 Programme. The vessels were to be based on ACASTA, with her endurance increased to between 4,000/4,500 miles. Another memo of 4.3.29 gave the comments of the C. in C.'s of the Mediterranean and Atlantic Fleets:
    (i) The vessels needed to be about 2 knots faster.
    (ii) The armament of the leader should not be less than that of the destroyer (a criticism of KEITH).
    (iii) The vessels should be of approximately the same dimensions as the ACASTAs.

THE LEADER: KEMPENFELT

The D.N.C. produced 4 sketch designs:
'A' : 325ft. 39,000 S.H.P. 37 knots (standard) on a displacement of 1,353 tons
'B' : 326ft. 41,500 S.H.P. 37 knots (standard) on a displacement of 1,385 tons
'C' : 328ft. 40,000 S.H.P. 37 knots (standard) on a displacement of 1,415 tons
'D' : 330ft. 39,000 S.H.P. 37 knots (standard) on a displacement of 1,450 tons

Design 'B' was based on AMAZON, design 'C' was an intermediate AMAZON/CODRINGTON and 'D' was based on CODRINGTON. The D.N.C. made the following comments:

    (i) Designs 'C' and 'D' were practical ships, with the desired propulsive efficiency.
    (ii) The extra armament would impose a weight penalty of 20 tons on the design.
    (iii) The vessels would be slightly longer.
    (iv) Additional accommodation to be provided by extending the deckhouse.
    (v) The fourth gun would be installed between the funnels.
    (vi) The speed required could be obtained with a two-boiler arrangement.

However, it was agreed on 22.4.29 that the 4th gun was to be carried in the same position as No. 3 gun in CODRINGTON and that the speed was to be the same as the destroyer. The proposed vessel was a repeat KEITH modified in the following ways:

(a) The vessel's length to be increased by 6ft to 318ft (reduced to 317¾' by 11.29) to accommodate the extra fuel.
(b) Breadth increased by 6 inches to 32ft 9 inches.
(c) Draft increased by 1" to 8 feet 7 inches.
(d) Standard displacement increased by 60 tons to 1,390 tons.
(e) Shaft horse power increased by 2,000 to 36,000.
(f) Speed increased by ½ knot at trial and deep displacement over KEITH.
(g) Oil fuel stowage increased by 70 tons to 450 tons.
(h) Additional 4.7" gun carried.
(i) Depth charges carried reduced by 9 to 6 charges and one rail.

CATEGORY: Weights in tons	KEITH 1928 Leader	KEMPENFELT 1929 Leader	
		5/29 DESIGN	11/29 DESIGN
General Equipment	106	110	112
Armament	104	122	120
Machinery	505	510	516
Hull	615	648	642
Displacement	1,330	1,390	1,390

The drawings of KEMPENFELT were approved on 14.11.29. Samuel White's tender of £228,638 dated 30.4.30 was accepted on 15.7.30. This tender price, excluding Admiralty supplied items, was made up of a hull price of £97,600, main and auxiliary machinery £123,210 and other machinery £7,828.

PROPOSALS FOR LEADERS ARMED WITH 6" GUNS

When KEMPENFELT's design was being discussed, sketch designs were prepared for a Flotilla leader of:
(a) 52,000 S.H.P. with four boilers to produce 34 knots on a deep displacement of 2,530 tons. Dimensions were 375' x 37' x 9¾'.
(b) 65,000 S.H.P. with five CODRINGTON type boilers to produce 33¾ knots on a deep displacement of 3,260 tons. Dimensions were 400' x 39' x 11'.

The designs were criticised on the grounds that "two super-imposed 6" guns forward and the resultant high bridge would involve much top hamper and if only one 6" gun were placed forward, difficulty would be experienced in making arrangements for the other three."

The designs were not progressed although KEMPENFELT was fitted with a Vickers Armstrong 5.1" gun on No. 2 mounting at Devonport during 6.32. The weapon, however, was not a success. It was felt that the 4.7" gun was still the standard weapon of the Japanese and was more adaptable. The Japanese at this time (1930) were regarded as the most likely enemy in any future war.

THE DESTROYERS: THE CRUSADERS

In early 1929, discussions were continuing on how to improve the endurance of the next group of destroyers. Stanley Goodall, the then head of the destroyer section, outlined the effects on the next destroyer design of increasing the endurance. Three sketches were drawn-up:
(A) An ACASTA, with an endurance of 4,000 miles at 15 knots, would be 10 feet longer, 50 tons heavier and would cost £6,500 more than the original design.
(B) An ACASTA, with an endurance of 4,500 miles at 15 knots, would be 20 feet longer, 110 tons heavier and would cost £20,000 more.
(C) Without increasing the length of the destroyer, an endurance of 4,000 miles could be achieved by a combination of increasing the depth by a foot, increasing the beam by between 3"-6", deep displacement by 100 tons and cost by £6,500. Speed, however, would be reduced by 0.3 knots in the standard condition and by 0.85 knots deep laden.

At the Sea Lords' meeting of 19.4.29, the following decisions on destroyer policy were reached:
    (i) Speed to be 32 knots in deep condition.
    (ii) Endurance to be 4,000 miles at 15 knots.
    (iii) 10' extra length per vessel was to be accepted.
    (iv) Leader to mount a 4th gun (deficient in KEITH).

The design of the 1929 destroyer was to be based on these criteria. However, the D.N.C. later reported that the endurance requirements could be met without increasing the length and beam of the proposed vessels. The general arrangement would be similar to that of ACASTA and BEAGLE.

The 1929 vessels would be fitted with Twin Speed Destroyer Sweeps, but not with ASDIC.

Accommodation for a Captain was to be provided in one ship of the group. The machinery of the vessels was to be up-rated to 36,000 S.H.P. and the vessels were consequently a ½ knot faster. Oil storage was to be 470 tons, instead of 380 tons of the BEAGLEs, giving the vessels an endurance of 200 miles more than the required 4,000 at 15 knots.

On 13.11.29, Board Minute 2656 approved the Legend of Particulars and drawings of the destroyers and Flotilla Leader of the 1929 Programme.

DETAIL	BEAGLE	1929 DESTROYER	KEITH	1929 LEADER
Length (PP)	312'	317¾'	312'	317¾'
Length (OA)	323'	329'	323'	329'
Standard Draft	8'6"	8'6"	8'6"	8'7"
S.H.P.	34,000	36,000	34,000	36,000
Speed (Deep) Knots	31½	32	31½	32
Oil Fuel (tons)	380	470	380	470
Weights (tons)				
General Equipment	100	104	106	112
Armament	122	126	104	120
Machinery	505	516	505	516
Hull	603	629	615	642
Margin	—	—	—	—
Standard Displacement	1,330	1,375	1,330	1,390

Their armament was the same as the BEAGLEs with the number of depth charges reduced from 15 to 6 in number.

When CYGNET was inclined at Buccleuch Dock on 30.1.32, her deep displacement was 1,914 tons, compared with her designed deep displacement of 1,865 tons — a growth of 49 tons. This was made up of the additional weight of 4.7" guns 4.7 tons, 3" gun and ammunition 4.3 tons, Torpedo Tubes 0.50 tons, modified Bridge design 5.6 tons, stiffening under 3" gun 1.8 tons, Director Tower 2.2 tons, Miscellaneous items 4.3 tons, re-assessment of judgement items 10.4 tons, oil fuel 4.0 tons and Machinery 7.0 tons — 44.2 tons in total. On sea trials CYGNET on a displacement of 1,448 tons reached a speed of 37.8 knots, compared with her design speed of $35\frac{1}{2}$ knots.

REDUCTION IN THE 1929 PROGRAMME

On 30.1.30 the revised 1929 Programme was announced consisting of a 6" gun cruiser, 1 Flotilla leader, 4 destroyers and 4 sloops. Half the projected flotilla was therefore cancelled. This was a political decision, as the London Naval Conference was in progress and the British Government had to take a lead in the reduction in armaments.

NAME	BUILDER	LAID DOWN	LAUNCHED	COMPLETED
KEMPENFELT	J. S. White	18.10.30	29.10.31	30. 5.32
COMET	Portsmouth D.Y.	12. 9.30	30. 9.31	2. 6.32
CRESCENT	Vickers Armstrongs	1.12.30	29. 9.31	15. 4.32
CRUSADER	Portsmouth D.Y.	12. 9.30	30. 9.31	2. 5.32
CYGNET	Vickers-Armstrongs	1.12.30	29. 9.30	1. 4.32

The five vessels of this group were awkward to operate within the Royal Navy's flotillas of nine destroyers and with the Royal Canadian Navy having a requirement to replace its obsolescent destroyers, the half flotilla was sold as a group to them between 1937/39.

KEMPENFELT spent 7 years in Home Waters, the Mediterranean and the Red Sea before her transfer to the Royal Canadian Navy on 19.10.39 as ASSINIBOINE.

KEMPENFELT/H.M.C.S. ASSINIBOINE (D18)

KEMPENFELT was completed on 30.5.32 as the leader of the 'C' class half flotilla and formed part of the 2nd D.F. of the Home Fleet. After operating from Rosyth for most of 1932, she left for the Mediterranean during 1.33 and operated there and in the Atlantic until 28.3.33. She then returned to home waters and after a brief courtesy visit to the Baltic ports, she returned to Devonport for alterations, which were completed during 1.34. KEMPENFELT and her half flotilla then participated in the Home Fleet's tour of the West Indies which was completed in 3.34. The remainder of 1934 was spent on a tour of Scandinavian ports, which was followed by operations in Scottish waters.

After a brief training cruise to the Mediterranean, the first eight months of 1935 were spent in home waters, but on 31.8.35, at the start of the Abyssinian crisis, she sailed for the Red Sea and service on the East Indies station until 4.36. In that month she returned to the U.K. and immediately refitted at Devonport until 6.36. Over the next six months she made a number of deployments to the Mediterranean and visited several Spanish ports evacuating British nationals from the civil war.

KEMPENFELT again refitted at Devonport between 12.36 and 10.4.37, visiting Spanish ports during the winter of 1937/38. After docking at Chatham in 5-6.38, she made another tour of Scandinavia in 7.38, and then returned to home waters for the remainder of 1938. KEMPENFELT re-commissioned in the Portsmouth Local Flotilla until the outbreak of war, when she immediately joined the 18th D.F. of the Channel Force, based at Portland, and was employed on escort and anti-submarine duties for the next month.

On her transfer to the Canadians on 19.10.39 as ASSINIBOINE she worked up and arrived at Halifax on 17.11.39, beginning escort duties the next day. She was then assigned to the Jamaican Force during 12.39, relieving SAGUENAY, and remained in the Caribbean until 6.40. The highlight of this period of duty was her capture of the German merchant ship HANNOVER in Mona Passage and her escort into Jamaica on 8.3.40 (HANNOVER subsequently became the first and short lived escort carrier AUDACITY).

In 6.40 ASSINIBOINE returned to Halifax, where she remained on local escort duties until 15.1.41, when she sailed for the U.K. to join the 10th Escort Group based at Greenock to undertake Atlantic escort duties for the next six months. On 28.2.41, ASSINIBOINE received survivors from the cargo vessel ANCHISES, which had been sunk by aircraft west of Ireland. On 5.4.41 she was damaged in a collision with the motor vessel LAIRDSWOOD and was under repair until 22.5.41. She was then assigned to the newly formed Newfoundland Command in 6.41 to provide mid-ocean escorts for Atlantic convoys. ASSINIBOINE and SAGUENAY provided part of the escort for the battleship PRINCE OF WALES at Placentia Bay during Anglo-American meetings and escorted the battleship as far as Iceland.

ASSINIBOINE was to continue on escort duties until 3.8.42, when whilst escorting convoy SC94 she rammed and sank U210. She was badly damaged in this encounter and was under repair and refit at Halifax from 29.8.42 to 20.12.42. ASSINIBOINE had barely worked up, when on 2.3.43, whilst she was on passage to Londonderry a U-boat was detected and ASSINIBOINE attacked with a pattern of depth charges that were set at too shallow a setting and she succeeded in badly damaging her stern. Repairs, undertaken at Liverpool, were completed between 7.3.43 and 13.7.43. She then joined Escort Group C1 of the Mid Ocean Escort Force until 4.44 when she returned to Canada to refit at Shelbourne Naval Station. Work was completed during 7.44. ASSINIBOINE then returned to European waters until the end of the war, serving with the 12th Escort Group at Londonderry, the 11th Escort Group and from 12.44 the 14th Escort Group at Liverpool. On the night of 11/12.8.44 SKEENA, RESTIGOUCHE, ALBRIGHTON and ASSINIBOINE were in action against German armed tawlers in Audierne Bay. Another mishap occurred on 14.2.45 when ASSINIBOINE collided with SS EMPIRE BOND in the Channel and was under repair until early 3.45.

In 6.45, ASSINIBOINE returned to Canada and a brief period of trooping, before a boiler room fire on 4.7.45 ended her service career. She paid off on 8.8.45 and was handed over to the War Assets Disposal Corporation after being declared surplus to requirements. She was sold for demolition at Baltimore, but on 10.11.45 whilst in tow for that port she broke her tow and was wrecked near East Point, Prince Edward Island. She was broken-up in situ in 1952, but a little of "Bones", as she was affectionately known in Canadian service, was still to be seen into the 1960s.

COMET/H.M.C.S. RESTIGOUCHE (HOO)

COMET was member of the 2nd D.F. of the Home Fleet between 6.32-7.34 after commissioning on 2.6.32. However, only a few weeks after commissioning, COMET collided with her sister CRESCENT at Chatham on 21.7.32. Repairs to COMET's propellers were completed at Chatham between 28.7-20.8.32.

After a refit and docking at Chatham between 20.7-3.9.34, COMET rejoined the 2nd D.F. until 4.36. COMET, like her sisters was detached to the Red Sea between 8.35-4.36. On her return to the U.K. she refitted at Sheerness between 23.4-29.6.36 and rejoined the 2nd D.F. for five months. During this period she served as a vessel of the Non-Intervention Patrol off the Biscay ports.

On 4.11.36, she reduced to a special complement, while consideration was being given to her transfer to Canada with her sister CRUSADER. She remained in this state until 29.12.36 when she started to operate as a plane-guard to the aircraft carrier GLORIOUS, in the Mediterranean. This service lasted until 4.37, when she returned to the U.K. for repairs at Portsmouth between 24.5-18.6.37. COMET then returned to her duties in the Mediterranean for the next year.

A refit and docking at Chatham between 26.5.38-20.8.38 then followed. The Canadians took over responsibility for the vessel at the start of her refit. She commissioned with a special complement from the R.C.N. on 11.6.38 as RESTIGOUCHE.

RESTIGOUCHE then worked up in U.K. waters and made passage to Esquimalt, where she finally arrived on 7.11.38. She was to remain on the west coast until ordered to leave for Halifax on 15.11.39. A highlight of this portion of her career was to escort the King and Queen during their Royal Tour of Canada.

When she reached Nova Scotia, she performed local escort duties from Halifax until 24.5.40, when she left for Plymouth. On arrival at Plymouth on 31.5.40, her after torpedo tubes were replaced by a 3" AA

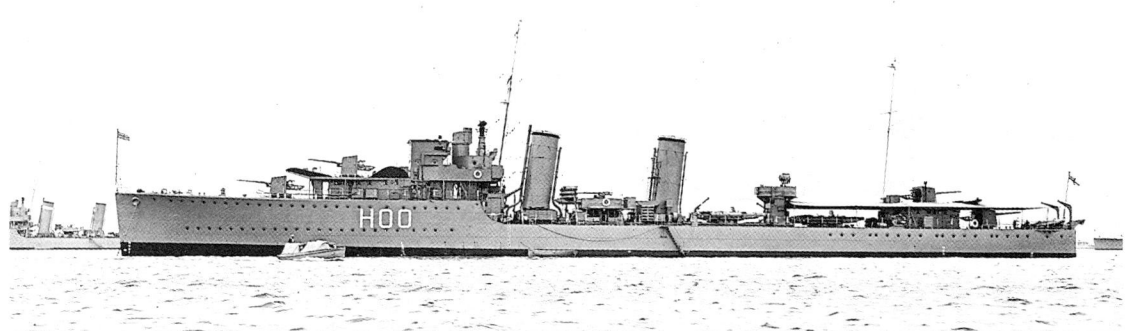
An early photograph of COMET, showing the separation of her bridge and the director control tower.

gun and the 2 pounder pom-pom by 0.5" machine guns. RESTIGOUCHE participated in the evacuation of troops from France and rescued survivors from FRASER in the Gironde on 25.6.40. She then operated on convoy duties with the Western Approaches Command until 8.40, when she returned to Canada to refit at Halifax until 10.40. She then undertook local duties until 1.41 when she returned to the U.K. to join the Western Approaches Command for 6 months.

RESTIGOUCHE was then allocated to the Newfoundland Command on 30.5.41 as a Mid Ocean escort until 4.43. On 4.8.41 whilst at Placentia Bay, she touched bottom and damaged her propellers and was under repair at St John's, Newfoundland and Halifax until 10.41. She did not remain on duty long, as on 13.12.41, whilst en-route to join ON44 she was badly damaged in a storm. Repairs were not completed at Greenock until 9.3.42.

In 4.43 RESTIGOUCHE joined Escort Group C4 for thirteen months on convoy escort duties, before being allocated to Escort Group 12 for the Normandy invasion. RESTIGOUCHE refitted on the Tyne from 8-12.43. After 'D' Day, she undertook Channel and Biscay patrols from Plymouth as part of Escort Group 12, before returning to Canada to refit during 9.44.

On the night of 7-8.6.44, she was missed by no less than 10 acoustic torpedoes in the Channel. On 5.7.44 when with QU'APPELLE, SASKATCHEWAN and SKEENA, RESTIGOUCHE participated in an action off Brest with two U-boats and three trawlers. Three days later she picked up survivors from U243 and was in action with three trawlers on 12.8.44. On 18.8.44, RESTIGOUCHE with CHAUDIERE, KOOTENAY and OTTAWA sank U621.

The refit that was begun at St John's was completed at Halifax and she then worked up at Bermuda, before returning to Halifax on 14.2.45 for local escort duties until 'VE' Day. Between 5.45 and 8.45, RESTIGOUCHE was employed on trooping duties from Newfoundland and was finally paid-off on 5.10.45. Sold to Foundation Maritime of Halifax, she was broken up during 1946.

RESTIGOUCHE in 1945. Note the Hedgehog in lieu of B gun, the close range AA weapons forward of the bridge and their non standard shields. (National Maritime Museum N6484)

CRESCENT on completion in 1932. Note the ready-use ammunition beside A gun.
(National Maritime Museum N9436)

CRESCENT/H.M.C.S. FRASER (H48)
CRESCENT commissioned on 21.4.32 as a member of the 2nd D.F. of the Home Fleet and served with this flotilla until 4.36. She had barely worked up when she was in collision with her sister COMET at Chatham on 21.7.32. She was under repair at Chatham until 27.8.32. Her service with the 2nd D.F. was punctuated by refits at Chatham 30.3-6.5.33 and 27.7-3.9.34. The highlight of her first commission was a deployment to the West Indies between 1-3.34. CRESCENT with her sisters was detached from the Home Fleet to the Red Sea and Indian Ocean between 9.35-4.36 at the height of the Abyssinian crisis.

On her return to the U.K. CRESCENT refitted at Sheerness 23.4-13.6.36. She then remained at Sheerness until 9.9.36, when she became the tender to the cruiser CARDIFF in reserve at the Nore, until 17.12.36. CRESCENT and her sister CYGNET had been sold to Canada at a combined cost of £400,000 on 20.10.36. A further refit at Canada's expense followed, which included the fitting of ASDIC. CRESCENT was finally taken over by the Canadians on 1.2.37 and commissioned into the Royal Canadian Navy as FRASER at Chatham on 17.2.37.

FRASER then made passage to the Pacific coast of Canada and finally arrived at Esquimalt on 3.5.37. She was then stationed on the west coast until 31.8.39 when she was ordered to the Canadian east coast and was passing through the Panama Canal on the outbreak of war. She was then based at Halifax from 15.9-11.39, coming under the operational control of the Royal Navy's America and West Indies Station. FRASER, however, remained based at Halifax over the winter of 1939/40, and joined the Jamaica Force for Caribbean patrols during 3.40.

On 26.5.40, she left Bermuda for the U.K. arriving at Plymouth eight days later, too late to participate in the Dunkirk evacuation. However, FRASER was immediately involved in the evacuation of allied troops from ports along the west coast of France.

On 25.6.40, whilst on passage in the Gironde Estuary after leaving St Jean de Luz, FRASER was in collision with the cruiser CALCUTTA. FRASER was cut in two forward of the bridge and sank immediately with the loss of 45 of her crew.

CRUSADER/H.M.C.S. OTTAWA (I) (H60)
CRUSADER commissioned as a vessel of the 2nd D.F. on 2.5.32 and remained with this flotilla for the next four years. She refitted at Portsmouth 30.7-4.9.34 and 27.4-30.5.36. She served in home waters until 9.35, when she joined the Mediterranean Fleet as a reinforcement during the Abyssinian crisis. A month later, with her sisters, she detached to the Red Sea and did not return to the U.K. until 4.36 when she refitted.

CRUSADER in the Solent during 7.32, shortly after completion. The separation of the bridge from the D.C.T. is clearly seen.

She then spent the next few months on miscellaneous duties and carried the C. in C. Home Fleet between Plymouth and Portsmouth on 20.7.36. CRUSADER spent 7-8.36 on evacuation and patrol duties off the Spanish Biscay coast. She then attended the battleship ROYAL OAK for torpedo trials, before beginning her duties as attendant destroyer to the carrier COURAGEOUS between 1.37-3.38. CRUSADER refitted at Portsmouth between 30.3-27.4.37 and 28.4-20.8.38 at Sheerness. During this latter refit, she was prepared for service with the Canadians and was renamed OTTAWA on commissioning on 15.6.38. After working up in the U.K., OTTAWA made passage to Esquimalt, where she arrived on 7.11.38.

She was to remain on the west coast of Canada for the next year and did not arrive at Halifax until 7.12.39. Although officially assigned to the America and West Indies Station, she remained as a local escort for eastbound convoys until 8.40. However, during 4.40 she was damaged in collision with the tug BANSURF and repairs to OTTAWA's stern were not completed for two months.

An early wartime photograph of OTTAWA showing both banks of torpedo tubes, one of which was removed during 10.40. She retains her original bridge.

On 27.8.40, she left Halifax for the Clyde where she was assigned to the 10th Escort Group and was based at Greenock until 6.41. On 25.9.40, she rescued 118 survivors from the merchant ships EURYMEDON and SULAIRIA, torpedoed when part of Convoy OB217. In 10.40, one of OTTAWA's sets of torpedo tubes was removed and a 3" AA gun fitted in lieu.

On the formation of the Newfoundland Escort Force in 6.41 she joined that command and was based at St. John's as a mid ocean escort for the next eleven months. OTTAWA transferred to Escort Group C4 during 5.42 and was a member of this group until her loss.

On 13.9.42, whilst escorting convoy ON 127, OTTAWA was twice torpedoed by U91 in the Gulf of St. Lawrence. She immediately broke in half and sank with the loss of the C.O., four other officers and 109 ratings. 69 survivors were picked up by the corvettes CELANDINE and ARVIDA.

CYGNET on 29.6.32 as a member of the Home Fleet's 2nd D.F.

CYGNET/H.M.C.S. ST. LAURENT (H83)

After commissioning on 9.4.32, CYGNET joined the 2nd D.F. of the Home Fleet until 7.34. During this time she was repaired at Devonport between 22.11.32-4.1.33, 29.3-6.5.33, 23.11.33-2.1.34 and received alterations and amendments, also at Devonport, between 20.7-24.8.33. She deployed to the West Indies, with the Home Fleet, between 1.34 and 3.34. On her return she again repaired at Devonport between 4.4-4.5.34, before re-fitting at Devonport between 25.7-31.8.34. On re-commissioning, she returned to the 2nd D.F. for service with the Home Fleet until detached to the Red Sea between 9.35-4.36 during the Abyssinian crisis.

CYGNET, after a refit at Devonport between 20.4-18.6.36, rejoined the 2nd D.F. on non-intervention patrol duties off the Biscay ports until 8.36, when she was replaced by the newly completed HUNTER. CYGNET entered reserve at the Nore on 30.9.36, where she remained until 15.12.36. She then started a refit prior to her transfer to the Canadians as had been arranged the previous October. The refit was completed on 13.2.37 and CYGNET recommissioned as H.M.C.S. ST. LAURENT four days later.

ST. LAURENT, on her arrival in Canadian waters, was stationed at Halifax for a year from 5.37 and then at Esquimalt until the outbreak of the war. She was then transferred to the east coast to undertake escort duties at Halifax, where she arrived on 15.9.39. ST. LAURENT remained on local escort duties for the next eight months and with FRASER escorted the detached squadron carrying gold bullion to Canada in 10.39.

On 24.5.40, she was ordered to the U.K. and after her arrival on 31.5.40 she was assigned to the Western Approaches Command and immediately participated in evacuation duties from ports along the western coast of France during 6.40.

On 2.7.40, she rescued 860 survivors from the liner ARANDORA STAR, which has been torpedoed and sunk that day in position 55° 20' N 10° 33' by U47 (Prien). Seamen were also rescued from the TITAN on 4.9.40 and the tanker CONCH and the A.M.C. FORFAR on 1.12.40. ST. LAURENT remained on escort duties with the Western Approaches Command until ordered to Canada during 2.41 to refit. She arrived at Halifax on 3.3.41. On completion of her refit she joined the Newfoundland Command as a mid ocean escort on 11.7.41 as part of 14th Escort Group until 12.41 and then as part of the Mid Ocean Escort Force until 3.43. She then joined Canadian Group C1 until 3.44, based at Londonderry. During this time, whilst escorting convoy ONS154, ST. LAURENT participated in the destruction of U356 on 27.12.42 with the corvettes CHILLIWACK, BATTLEFORD and NAPANEE. Fifteen months later, as part of the Escort Group C1, ST. LAURENT with the destroyer FORESTER and the Canadian frigates OWEN SOUND and SWANSEA, sank U845 on 10.3.44 in the North Atlantic when escorting convoy SC154.

During this period she refitted at Halifax between 4.42-9.8.42 and at Dartmouth Nova Scotia between 17.8-12.43.

In 5.44, she transferred to the 11th Escort Group for invasion duties and remained with the group until 11.44. She had a narrow escape on 8.8.44 when she was near-missed by a glider bomb from a German aircraft in the Bay of Biscay. However, by this time ST. LAURENT needed major repairs, which were undertaken at Shelbourne Naval Station between 11.44-20.3.45.

ST. LAURENT remained in Canadian waters until after V.E. Day and then spent several months on trooping duties, before being paid off into Reserve at Sydney on 10.10.45. She was subsequently sold to the International Iron and Metal Co. of Hamilton and was scrapped during 1947 at Levis, P.Q.

THE 1930 PROGRAMME

THE LEADER: DUNCAN

DUNCAN was generally a repeat of KEMPENFELT. There was very little discussion over her design. She varied from KEMPENFELT only in her displacement which was 10 tons greater and the composition of her weights which were slightly different. The armament was unchanged.

Tenders were sought on 24.9.30 and DUNCAN was ordered from Portsmouth Dockyard on 2.2.31, with machinery being supplied by Beardmore's at a cost of £139,768.

THE DESTROYERS: THE DEFENDERS

On 28.5.30 it was agreed that the destroyers of the 1930 New Construction Programme were to be repeats of the CRUSADER type. Two months later, it was decided to fit the vessels with ASDIC and not T.S.D.S.

Discussions centred on the need for high angle armament to be separate from the main armament. No consideration seems to have been given to a high angle/low angle 4.7" mounting at this time. However, some limited modifications were made in the design of the DEFENDERs compared with that of CRUSADER:
(i) The 2 pounder pom-pom was to be replaced by a 3" high angle weapon and two 0.5" machine guns were added on the forecastle deck.
(ii) The vessels were to have the same number of depth charges as the BEAGLEs — one rail, two throwers and 20 charges.
(iii) ASDIC to be provided.
(iv) The machinery was slightly lighter than that in the CRUSADERs, but provided the same power.

The Board approved the legend and drawings on 7.10.30 with orders being placed on 2.2.31 for the eight destroyers and the contract prices were as follows:

Name	Builder	Hull	Main Machinery	Auxiliary Machinery	Total
DEFENDER	Vickers-Armstrongs	£99,685	£116,405	£7,889	£223,979
DIAMOND		£99,215	£116,405	£7,889	£223,509
DARING	Thornycroft	£94,400	£123,600	£7,536	£225,536
DECOY		£94,100	£123,600	£7,536	£225,236
DIANA	Palmers	£99,300	£122,000	£8,202	£229,502
DUCHESS		£99,165	£122,000	£8,202	£229,367
DAINTY	Fairfield	£99,840	£121,955	£7,583	£229,378
DELIGHT		£99,840	£121,955	£7,583	£229,378

On trials, DEFENDER on 1,440 tons trial displacement and at 35,910 S.H.P. achieved a speed of 37.2 knots (design speed was 35.5 knots) on 14.9.32. DARING, when inclined was 1,897 tons deep loaded, compared with a designed deep displacement of 1,884 tons.

Name	Laid down	Launched	Completed
DUNCAN	25. 9.31	7. 7.32	31. 3.33
DAINTY	20. 4.31	3. 5.32	22.12.32
DARING	18. 6.31	7. 4.32	25.11.32
DECOY	25. 6.31	7. 6.32	17. 1.33
DEFENDER	22. 6.31	7. 4.32	31.10.32
DELIGHT	22. 4.31	2. 6.32	31. 1.33
DIAMOND	29. 9.31	8. 4.32	3.11.32
DIANA	12. 6.31	16. 6.32	21.12.32
DUCHESS	12. 6.31	19. 7.32	27. 1.33

DUNCAN on completion in 4.33 as leader of the First D.F. of the Mediterranean Fleet.

DUNCAN (D99)

DUNCAN arrived at Malta on 7.5.33 and became leader of the First D.F. on this station for the next fifteen months with a detachment to the Red Sea between 9-11.33. After refitting at Portsmouth between 3.9-23.10.34, DUNCAN led her flotilla on passage to the Far East, arriving at Hong Kong on 3.1.35. She was to remain on the China Station as leader of the 8th D.F., renumbered the 21st in 4.39, until the outbreak of war.

Highlights of her career in the Far East were a visit to Kobe, Japan's foremost naval base, between 26.4-11.5.36 and tours of Singapore, Malaya and Thailand between 12.35-1.36, the Philippines and Dutch East Indies a year later. Another tour of Singapore, Saigon and Manila was completed between 1-3.38.

On 14.12.36 DUNCAN was taken in hand at Hong Kong for damage sustained whilst undertaking oiling at sea trials. Repairs were completed on 4.1.37. A few months later, between 5-21.8.37, DUNCAN lay at Shanghai during the Japanese invasion and with the sloop FALMOUTH she evacuated British women and children from Shanghai to Woosang, where they arrived on 21.8.37.

At 10.10 hours on 28.10.38, whilst lying at Sharp Peak, Foo Chow, she was in collision with the Greek steamer PIPINA. Collision repairs and a refit at Hong Kong were completed between 31.10.38-14.1.39. Early in 7.39, DUNCAN was hit by a high speed target at the fleet anchorage of Wei Hai Wei; after four days of temporary repairs, and after being escorted as far as Amoy by DECOY, DUNCAN was repaired at the Taikoo dockyard for damage to her bow. These repairs were completed on 7.8.39.

On the outbreak of war DUNCAN with DAINTY, DIANA and DARING (from Singapore) made passage to the Mediterranean and arrived at Alexandria on 30.9.39. The whole class were in a poor material condition and during their stay in the Mediterranean, repairs were undertaken to the many defects that had shown up. They then undertook contraband control work.

On 6.12.39, DUNCAN left Gibraltar as escort for the battleship BARHAM with DUCHESS. DUNCAN arrived in the Clyde on 12.12.39 and was immediately assigned to the 3rd D.F.

However, on 17.1.40 in position 58°27'N 02°00'W, she was in collision with a vessel from convoy ON8. She was towed into Invergordon by IMPULSIVE the next day, with EXMOUTH (lost three days later) as screen. After temporary repairs at Invergordon, she was towed from there by the tugs ST MELLONS and NORMAN, escorted by DIANA, BRAZEN and BOREAS to Grangemouth on 15.2.40. Repairs were not completed until 22.7.40.

DUNCAN then returned to the 3rd D.F. at Scapa until 10.40. She left the Clyde as escort for the aircraft-carrier ARK ROYAL on passage to the Mediterranean and joined the 13th D.F., based at Gibraltar until 3.41. During this period she participated in the attack on Cagliari, Sardinia on 11.11.40 and a supply operation to Malta from 25.11.40. On 1.1.41 with other flotilla members, she intercepted a Vichy French convoy off Oran and six days later participated in operation "EXCESS" — the passage of convoys east and west across the Mediterranean. On 31.1.41 DUNCAN, as part of Force 'H', participated in the bombardment of Genoa (Operation "RESULT").

On 10.3.41 she sailed to Freetown as escort for the battle-cruiser REPULSE and the aircraft carrier FURIOUS, conveying aircraft to Takoradi. She remained on the West African Command at Freetown until recalled to Gibraltar to participate in Operation "SUBSTANCE" during 7.41. She remained at Gibraltar with the 13th D.F. until 11.41, participating in operation "HALBERD" — the passage of a replenishment convoy to Malta between 24-30.9.41.

DUNCAN returned to the UK to refit at Sheerness between 16.11.41-23.1.42. She arrived at Greenock four days later with weather damage, which necessitated repairs until 4.2.42 when she sailed to Gibraltar to rejoin the 13th D.F. She participated in two flying-off operations to Malta during 3.42.

In 4.42 she was sent as a reinforcement for the Eastern Fleet's 22nd D.F. and arrived at Kilindini on 22.5.42. She escorted with the cruiser DEVONSHIRE a convoy for the occupation of Diego Suarez, Madagascar on 1.6.42. DUNCAN continued to operate with the 22nd D.F. until 9.42. She arrived at Greenock on 16.11.42 after escorting the battleship ROYAL SOVEREIGN via South Africa.

DUNCAN herself refitted at Tilbury, which included re-tubing her boilers and conversion to an A/S destroyer, between 24.11.42-18.3.43. After working-up at Tobermory she joined the Western Approaches Command and Escort Group B7 until 11.43.

Between 29.4-5.5.43, DUNCAN participated in the successful counter attacks against the U-boats which attacked ONS5 off Cape Farewell, when four U-boats were sunk. On 16.10.43, she picked up 15 survivors from U470, which had been sunk by a Liberator aircraft and 13 days later she and the destroyer VIDETTE and the corvette SUNFLOWER sank U282 in position 55°28'N 31°57'W whilst protecting convoy ON208.

DUNCAN's material condition was causing concern and structural and machinery repairs were undertaken at the North Woolwich yard of Harland and Wolff between 12.11.43-17.5.44. After working up at Tobermory between 21-29.5.44, DUNCAN returned to escort duties as a private ship with the 14th Escort Group of the Western Approaches Command until 4.45. At this time she joined the Greenock Coastal Escort Pool. She was not to remain active much longer as on 13.5.45 she was ordered to reduce to Category 'C' Reserve, one of the few inter-war destroyers to be on active service on the first and last days of the European War.

DUNCAN arrived at Barrow on 9.6.45 and was relegated to Category 'C' Reserve ten days later. However, as she was leaking approximately five tons of water a day due to defects, immediate disposal was sought. She was approved for scrapping on 8.7.45 and immediately turned over to BISCO for subsequent scrapping at Barrow by T. W. Ward Ltd, but demolition was not completed until 1949.

A dilapidated DUNCAN laid-up at Barrow on 12.8.48. She had been allocated to BISCO three years before, but demolition was not completed for another year.
(Ken Royall)

DAINTY (H53)

DAINTY spent her first commission with the First D.F. on the Mediterranean Station between 2.33 and 8.34; spending 10-11.33 on a tour of the Red Sea and Persian Gulf. Refitted at Portsmouth between 3.9-23.10.34, DAINTY commissioned on the latter day for the China Station and made passage to Hong Kong, where she arrived on 3.1.35.

DAINTY remained on station at Hong Kong until the outbreak of war, with the 8th D.F. until 4.39, when the flotilla was renumbered the 21st. She was detached to the Red Sea between 30.9.35-6.36 and after refitting and docking at Hong Kong between 21.9-15.10.36, she undertook anti-piracy patrols from Hong Kong. On 21.1.37 the steamer HSIN PEKING grounded on the Nemesis Rock off Ningpo. DAINTY put an anti-piracy guard on board and the vessel was refloated and proceeded on her journey. A tour of Singapore, Sarawak and the Philippines was undertaken between 1-3.38.

DAINTY on commissioning. Note that the bridge and director have been plated together.
(Strathclyde Regional Archives)

On 3.9.39 DAINTY with DUNCAN and DIANA proceeded to the Mediterranean, where they arrived on 30.9.39 to spend the next month on patrol duties, before refitting at Malta between 8-30.12.39. She was then attached to Force 'G' of the South Atlantic Command as part of the 2nd Destroyer Division and remained on station at Freetown until 4.40. She then returned to the Mediterranean, where she refitted at Malta between 21.4-2.6.40 before joining the 10th D.F. until her loss.

DAINTY participated in the early sweeps against Italian forces, participating in the rescue of over 400 survivors of the torpedoed cruiser CALYPSO, sunk whilst in company with the fleet south of Crete on 12.6.40. Between 27.6.40 and 29.6.40 DAINTY and the destroyer ILEX engaged no less than three Italian submarines, the CONSOLE GENERALE LUIZZI being scuttled after a gun duel on 27.6.40, the UEBI SCEBELI sunk by depth charges two days later, and on 29.6.40 SALPA was severely shaken in a depth charge attack. All the attacks occurred west of Crete. Between 20.7-20.8.40, DAINTY undertook escort duties for the arrival of the Australian Expeditionary Force in Egypt. Further escort and patrol duties kept DAINTY busy until 12.40, when she operated for a month on the flank of the army advancing into Libya, attempting to prevent seaborne supplies from reaching the Italian forces in the desert. On 31.12.40 DAINTY captured two 250 ton schooners attempting to run the blockade into Bardia.

In late January 1941 DAINTY, when based at Suda Bay, towed in the damaged tanker DESMOULEA under heavy air attack, after the latter had been torpedoed off eastern Crete.

DAINTY was to survive barely three weeks longer, because after returning to her duties with the Inshore Squadron she was caught off Tobruk at dusk on 24.2.41 by German and Italian dive bombers and hit aft by a 1,000lb bomb. The resulting fuel fire rapidly got out of control, the after magazine blew up, and she sank. 2 officers and 14 ratings were killed and another 18 ratings were wounded. She lies in 32° 04' N 24° 04' E.

DARING (H16)

After her first commission with the First D.F. of the Mediterranean Fleet between 1.33-8.34, and including a deployment to the Persian Gulf during 9-10.33 DARING refitted at Sheerness between 3.9-24.10.34 for service in the Far East. Her commander between 4-12.34 was Lord Louis Mountbatten who took her to Singapore in 12.34, but then transferred to WISHART.

DARING was to be based at Hong Kong as part of the China Squadron until 9.39, initially with the 8th D.F. and then the 21st D.F. The whole flotilla left Singapore before the commencement of hostilities, but DARING was retained in the Red Sea for escort and patrol duties until 11.39. This was followed by a period for docking and repairs at Malta between 25.11-20.12.39.

She finally arrived at Belfast on 7.1.40 as escort for the Union Castle liner DUNNOTTAR CASTLE—the first time she had been in the U.K. since 12.34. After a further period for repairs at Portsmouth until 25.1.40, she then left on passage for Scapa via the Clyde. She arrived at Scapa on 10.2.40 for service with the 3rd D.F.

She was to survive a mere eight days, as on 18.2.40, whilst escorting convoy HN12 from Norway she was torpedoed by U23 (Kretschmer) in position 58°40'N 01°40'W. Her stern was blown off and she capsized and sank within minutes. Only five ratings were picked up by the submarine THISTLE which surfaced after witnessing the attack. DARING's C.O., Cdr S.A. Cooper, 8 other officers and 148 ratings were lost. The vessel lies in position 58° 39' N 01° 40' W.

A pre-war view of DARING which joined the Home Fleet's 3rd D.F. a mere 8 days before her sinking by U23 on 18.2.40. There were only five survivors.

DECOY/H.M.C.S. KOOTENAY (H75)

After working up, DECOY joined the First D.F. in the Mediterranean during 4.33. She was detached to the Persian Gulf between 9-11.33, and then she returned to Malta for the fitting of new torpedo tubes between 7.11-1.12.33. DECOY returned to the U.K. during 8.34 and refitted at Devonport between 3.9-20.10.34 before making passage to Hong Kong between 1.11.34-3.1.35.

DECOY was based in the Far East until the outbreak of war, first with the 8th D.F. and later the 21st D.F. She was detached to the Red Sea from 11.35 for the duration of the Abyssinian crisis and was based on Aden and Perim. Before returning to the Far East, she visited Mombasa and other East African ports during 6-7.36. DECOY undertook local patrols at Hong Kong and along the China coast and refitted at Hong Kong between 1-31.10.36 and had to be further repaired and fumigated at Hong Kong between 13.4-29.5.37 after a tour of South-East Asia between 1-3.37. In 8.38 she carried to Tsing-Tao the representatives apologising for incidents when liberty-men insulted the Japanese Flag.

During 9.39 the flotilla transferred to the Mediterranean, where its vessels undertook contraband control work. However, on 2.12.39 when DECOY was refitting at Malta, it was discovered she had corrosion of the bulkheads between the engine and boiler-room, rendering her unseaworthy. Finally on 18.12.39 DECOY was taken in hand for 4½ weeks of continuous overtime work on feed-pump defects, replacement of funnels and a hull survey.

Despite the C. in C. Mediterranean drawing attention to the unsuitability of these destroyers for the South Atlantic as they would need constant Dockyard attention, DECOY was transferred to that Station during 1.40. She joined the 20th D.F. based at Freetown and remained on patrol and escort duties until recalled to the Mediterranean during 5.40, to join the 10th.

On 12.5.40 DECOY and DEFENDER met the Australian troop convoy US2 in the Red Sea and escorted it to Suez. On the night of the 20-21.6.40, DECOY with DAINTY, HASTY and the R.A.N. STUART bombarded Bardia. Further patrol and escort duties then followed, including the escorting of the cruisers ORION and

DECOY on her return from the Far East in 1939. She was later to see service in the South Atlantic, Mediterranean and the Indian Ocean on fleet duties..

SYDNEY bombarding Scarpanto on 3-4.9.40. However, on 13.11.40 DECOY was hit by a bomb during an air-raid at Alexandria. The bomb damaged No. 4 4.7" mounting and the upper and main decks and the ship's bottom was penetrated between stations 146 and 153. DECOY transferred to Malta for permanent repairs, which were completed on 1.2.41, after further minor bomb damage was received on 19.1.41.

DECOY and HEREWARD transported and landed 200 commandos on the island of Castellorizo, east of Rhodes on 25.2.41. However, the commandos were overwhelmed two days later, so no evacuation was required.

DECOY was involved in the Greek and Cretan evacuations, being slightly damaged by a near miss on 31.5.41. She was a regular on the Tobruk run between 7-11.41, again being near missed whilst returning from Tobruk to Alexandria on 9.7.41. Between 20.12.41-8.2.42 DECOY was under repair for collision damage at Malta. On completion of these repairs, she joined the escort of convoy ME10 from Malta to Alexandria.

DECOY had been assigned to the 2nd D.F. of the Eastern Fleet and by 31.3.42 she was part of the escort of Force B — the slow vessels of the Eastern Fleet, south of Ceylon. The Eastern Fleet was almost devoid of air cover and had to avoid action with the triumphant Japanese carriers.

In 9.42, when overdue for refit, she started her passage back to the U.K. but operated from Freetown during 9-10.42. DECOY arrived at Greenock on 29.10.42 — the first time she had been home since 11.34. DECOY refitted at the old Palmers yard on the Tyne between 3.11.42 and 12.4.43, when she commissioned as H.M.C.S. KOOTENAY after transferring to the Canadians on 1.3.43 (her transfer to the Royal Canadian Navy as a gift was made permanent on 15.6.43).

DECOY, showing many war modifications, became the Canadian KOOTENAY on 1.3.43 and served on the Atlantic 'run' for the remainder of the war.

After working-up at Tobermory, KOOTENAY joined Escort Group C5, as part of the Mid Ocean Escort Force until 10.43. She was then under repair at Halifax until 12.43. Further ocean escort duties followed until KOOTENAY was assigned to 'D' Day duties and performed escort duties in the Channel and the Bay of Biscay until 9.44. During this period she participated in the sinking of U678 south of Brighton on 6.7.44, U621 off La Rochelle on 18.8.44 and U984 west of Brest two days later.

KOOTENAY refitted at Shelbourne Naval Dockyard between 2.10.44-27.2.45 and returned to the U.K. to work up at Tobermory and then operated from Plymouth until V.E. Day. KOOTENAY was utilised as a troop transport between Newfoundland and Quebec City before finally paying off into Reserve at Sydney, Nova Scotia, on 26.10.45. She was sold for demolition during 1946 to International Iron and Metal, Hamilton.

DEFENDER running trials during 10.32. (V.S.E.L.)

DEFENDER (H07)

After commissioning on 8.11.32, DEFENDER joined the First D.F. in the Mediterranean from 1.33-8.34 and acted as leader until DUNCAN came on station during 5.33. The flotilla toured the Red Sea and Persian Gulf between 9-11.33.

DEFENDER returned to the U.K. during 8.34 and immediately refitted at Devonport 3.9-23.10.34, before making passage to the Far East between 11.34 and 1.35, when she joined the 8th D.F., later the 21st, at Hong Kong until 8.39. She was detached to the Red Sea between 11.35-6.36 for the Abyssinian crisis followed by a month cruising along the East African coast. During 1938/39, DEFENDER had problems with her boilers and superheaters, the boilers being re-tubed at Singapore 5.11.38-26.1.39, and her superheaters repaired at Hong Kong 31.1-14.3.39.

Lying at Singapore at the end of 8.39, DEFENDER immediately made passage with her sisters to the Mediterranean, arriving at Alexandria on 19.9.39. She remained with the 21st D.F. in the Mediterranean until 1.40 where she served on contraband control patrols. During 1.40, she undertook patrols off Portugal based at Gibraltar. Between 18.2-18.4.40, she was based at Freetown on escort duties returning to Gibraltar as escort for the cruiser NEPTUNE on 23.4.40.

DEFENDER rejoined the Mediterranean Fleet as a unit of the 10th D.F. in 5.40 until her loss. She was soon in action, as 16 days after the opening of hostilities on 11.6.40, she and DAINTY hunted and sank the Italian submarine CONSOLE GENERALE LUIZZI south east of Crete. DEFENDER participated in the action off Calabria on 9.7.40, the Cape Spartivento action on 27.11.40 and the battle of Cape Matapan on the night of 28/29.3.41. Beside these actions, DEFENDER undertook escort duties to Malta, Greece and was present at the evacuation of British forces from Crete during 5.41. Previously, DEFENDER had refitted at Malta between 4.2-19.3.41.

DEFENDER was lost whilst escorting a convoy to the besieged port of Tobruk. On 11.7.41 she was badly damaged by an aircraft bomb which exploded under the ship just aft of the forward bulkhead of the engine room. At 05.55 hours she signalled "Engine room flooded. Back broken but have every hope of saving the ship".

At 06.46 hours, the Australian destroyer VENDETTA signalled "Have DEFENDER in tow. Request max. protection". At 07.15 hours VENDETTA signalled of DEFENDER "Heavily listed. Skeleton crew on board". DEFENDER sank off Sidi Barrani at 11.15 hours on the same day. There were no casualties.

DEFENDER settling with her back broken, after being damaged by air attack off Sidi Barrani on the morning of 11.7.41.
(Imperial War Museum HU52865)

H.M.S. DELIGHT (H38)

After work up at Portland, DELIGHT was used for trials with the new Mk IX torpedo until 29.4.33. She then joined the Mediterranean Fleet's First D.F. until 8.34. She was, however, on detached service in the Persian Gulf between 9.33 and 11.33.

A pre-war view of DELIGHT which became an unnecessary loss to air attack when sunk 20 miles S.S.W. of Portland Bill on 29.7.40. She had sailed from Portland in daylight contrary to standing orders.

(Strathclyde Regional Archives)

On her return to the U.K. she refitted at Portsmouth between 3.9-25.10.34, before her passage to the China Station and service with 8th D.F. from 1.35. She remained there, with periodic refits at Hong Kong, until the war; between 9.35 and 11.35 she was detached to Aden for the Abyssinian crisis and a year later conducted 2 months of anti piracy patrols from Hong Kong.

DELIGHT left the China Station on 29.8.39 for Aden where she arrived on 14.9.39 and Alexandria later. She operated with her flotilla in the Mediterranean for the next three months. She finally arrived at Portsmouth on 30.12.39 after 5 years absence from the U.K. She then refitted at Portsmouth until 27.1.40, joining the Home Fleet's 3rd D.F. on 7.2.40.

A period of arduous service then followed with DELIGHT being damaged by heavy weather on 8.4.40 at the start of the Norwegian Campaign. On 7.6.40 she narrowly missed destruction by the German GNEISENAU and SCHARNHORST when escorting a troop convoy from Harstad to the U.K. On the 13th of that month she assisted the A.M.C. SCOTSTOUN which had been torpedoed by U.25 in position 57°00′N 09°57′W. DELIGHT had her super heaters retubed at Rosyth between 21.6-24.7.40. She was then ordered to the Clyde and whilst on passage to this destination she was lost.

Contrary to standing orders, DELIGHT left Portland in daylight on 29.7.40 and was picked up by a "FREYA" radar at Cherbourg when 20 miles S.S.W. of Portland Bill at 18.35 hours. She was subsequently attacked by 16 aircraft in position 50°12′N 02°17′W.

One bomb hit DELIGHT a glancing blow on her forecastle and she caught fire. The fire which started in the low power room was aided by an escape of oil fuel and the fracture of the main steam pipe. At 21.30 hours a large explosion occurred in the forward part of the ship and she sank shortly afterwards. 6 ratings were killed in this wholly unnecessary loss which came at a critical time for the Royal Navy.

DIAMOND (H22)

Commissioned on 8.11.32, DIAMOND, after working up at Portland, joined the First D.F. in the Mediterranean the following month until 8.34. She was detached to the Red Sea and the Persian Gulf during 9-11.33.

She refitted at Devonport between 3.9-27.10.34 and then with her sisters made passage to the China Station 11.34-1.35, where she joined the 8th D.F. Based at Hong Kong, DIAMOND remained on that Station until the outbreak of war.

On 7.8.39, she started a refit at Singapore, which continued until 11.39. DIAMOND did not leave Singapore until 4.12.39 and arrived at Malta fifteen days later. She did not remain in the Mediterranean long as the next month she was allocated to the South Atlantic Station, sailing from Malta on 8.1.40. She joined the 20th Destroyer Division at Freetown on 15.1.40 and undertook escort duties until 4.40.

DIAMOND running trials before commissioning. The director tower is now part of the bridge structure.
(V.S.E.L.)

DIAMOND returned to the Mediterranean that same month and after a short refit at Malta she joined the newly formed 10th D.F., which consisted of the 20th Division and five Australian destroyers. DIAMOND was soon in action, receiving minor damage by bombing off Malta on 11.6.40 and again six days later. She was involved in many convoy actions and sweeps, bombarding the Italian seaplane base at Bomba on 23.8.40; the escorting of convoy MB8 during Operation 'COLLAR' in 11.40 and the convoy operation 'EXCESS' in 1.41.

She was involved in the evacuation of British and Commonwealth forces from Greece and on 27.4.41 the transport SLAMAT was attacked whilst loading troops and was sunk. DIAMOND and the destroyer WRYNECK picked up 700 survivors and started for Crete. Shortly afterwards both ships were sunk south of Morea by air attack. Only one officer, 41 ratings and 8 soldiers survived from the three vessels. Her Commanding Officer Lt. Cdr. P.A. Cartwright, 7 other officers and 141 ratings from DIAMOND were listed as missing presumed killed. The vessel lies in position 36°30'N 23°34'E.

DIANA/H.M.C.S. MARGAREE (H49)

Commissioned at Chatham on 29.12.32, DIANA was a unit of the First D.F. of the Mediterranean Fleet between 1.33-8.34, being detached to the Red Sea and the Persian Gulf between 9-11.33. She refitted at Sheerness between 3.9-23.10.34 on her return to the U.K.

After completing her passage to the Far East during 1.35, DIANA served in these waters until 9.39. However, she had several detachments to the Red Sea between 9.35-5.36, during the Abyssinian crisis, at Bombay between 5-6.36, East Africa 6-7.36 and finally returned to her base, Hong Kong on 7.8.36.

A fine portrait of DIANA shortly after commissioning. She served in Mediterranean waters during 1939-40 before becoming the Canadian MARGAREE on 6.9.40.

There were two unusual occurrences during her period in the Far East, the first being on 30.9.37, when the prompt arrival of a party from DIANA on board the vessel FAUSANG at Swatow had a sobering effect on the "somewhat easily excited coolies". A few months later, DIANA performed a special service at Turnabout Island near Amoy to investigate the lighthouse situated there and the extinguished light. The lighthouse had been attacked by pirates.

On the outbreak of war DIANA and her sisters immediately returned from the China Station to the Mediterranean, where she spent 11.39 at Malta rectifying defects that had been caused by corrosion. She then undertook contraband patrol duties. She returned to the U.K. during 12.39 and joined the Home Fleet's 3rd D.F. and after participating in the Norwegian Campaign, she commenced a refit in the Royal Albert Dock in London during 7.40.

On 6.9.40 in the Royal Albert Dock, DIANA was commissioned into the Royal Canadian Navy as the MARGAREE. She was a replacement for the FRASER lost earlier and many of her crew came from that vessel. After working-up, MARGAREE left Londonderry for Canada with a five ship convoy OL8. 5 days later, on 22.10.40, she was lost in collision in the North Atlantic with the freighter PORT FAIRY with the loss of 142 of her crew. Six officers and 28 ratings were picked up by the PORT FAIRY.

DUCHESS served on the China Station between 10.34-9.39, but never got back to the U.K., being lost in collision with the battleship BARHAM off the Mull of Kintyre on 10.12.39.

DUCHESS (H64)

On 12.4.33, DUCHESS left the U.K. for the Mediterranean, where she served with the First D.F. until 8.34. She toured the Red Sea and Persian Gulf between 9-11.33. Repairs to her superheaters kept her at Malta between 18.12.33-6.1.34.

On her return to the U.K., she refitted at Chatham between 3.9-23.10.34, re-commissioning on 25.10.34 and made passage to China between 11.34 and 1.35, joining the 8th, later 21st, D.F. at Hong Kong. She was to serve with this Flotilla until the outbreak of war. She was never to return to the U.K.

Her service, especially her first commission between 1935/37, was to be varied, as she served in the Red Sea between 9-11.35 during the height of the Abyssinian crisis. 4-5.36 saw her on a courtesy visit to Japan, which was followed by anti-piracy patrols based on Hong Kong between 8-10.36. Singapore, Malaya and the Philippines were visited between 1-3.37. Anti-piracy patrols were again undertaken between 5-8.37.

On 2.9.37, whilst lying at Hong Kong, DUCHESS's stern was crushed when, in a typhoon, a merchant ship dragged her anchors and hit her. She was repaired and refitted at Hong Kong between 4.9-14.10.37. She then operated off the Chinese coast during 11-12.37. Normal duties then followed and between 1-3.39, she visited the Dutch East Indies, Singapore and Malacca.

DUCHESS left Hong Kong for the Mediterranean on 28.8.39, transiting via Singapore, Colombo, Alexandria and arrived at Malta on 12.10.39. She then remained in the Mediterranean for the next two months before leaving Gibraltar on 6.12.39 as escort for the battleship BARHAM returning to the U.K. However, on 10.12.39 she was in collision with the BARHAM in thick fog west of the Mull of Kintyre in position 55° 21' 52 N, 06° 02' 39" W. She sank with the loss of 124 officers and men.

THE 1931 PROGRAMME
THE FLOTILLA LEADER: EXMOUTH

The design of this vessel took over 18 months to finalise from being initiated by a minute to the D.N.C. from the Controller on 3.10.30:

"Provisionally agreed that the leader to be included in the 1931 programme should be of the leader, not the destroyer class in order that the additional accommodation required for the increased complement carried by the leaders should be provided."

It was anticipated that the vessel would be between 100/200 tons larger and that she should be slightly faster than the destroyers of her flotilla.

Progress was slow and it was not until the Controller's Conference of 9.4.31 that further parameters were settled. It was agreed then that the vessel was to be generally similar to CODRINGTON with increased endurance, but have a speed of ½ knot more than the E class destroyers.

However, on 28.5.31 the Controller directed the D.N.C. to prepare a three-boiler arrangement for the proposed vessel, and this was to complicate matters for the next year. The D.N.C. on 11.6.31 stated that decisions were required on the speed and shaft horsepower of the vessel, the number of boiler rooms, whether the vessel would deploy a low angle/high angle armament and be fitted with T.S.D.S. The D.N.C. commented that an increase of 3,000 S.H.P. would be advantageous for flotilla work and that a three boiler design should now be adopted as standard practice for all future leaders and destroyers. Furthermore, the provision of a fifth 4.7" gun would involve an increase in length and displacement; no high angle 60° gun for combined H.A./L.A. was recommended, but one could be substituted for a low angle mounting or for the 3" gun.

"Two 3" H.A. guns are shown in the design on the same platform, but it is not considered that the two guns thus mounted are sufficiently superior to one on the centre line to justify the additional weight. T.S.D.S. to be fitted."

On 11.7.31, the Controller made the following decisions:
(i) 3 boiler-room design is adopted.
(ii) The 4.7" H.A. to be positioned on a raised platform in lieu of the 3".
    The 3" to be carried until the 4.7" mounting becomes available.
(iii) 200 instead of 150 rounds per gun to be provided.
(iv) Two machine guns to be carried.
(v) Three months provisions and 4 months naval stores to be carried.

The D.N.C. then sketched out two designs:

Design (A) with three boiler rooms and one engine room. Dimensions 340' x 33¾' x 8¾' on a standard displacement of 1,505 tons. S.H.P. 38,000. Speed 36 knots. Oil fuel 490 tons.

Design (B) with three boiler rooms and two engine rooms. Dimensions 360' x 35' x 8⅔' on a standard displacement of 1,615 tons. S.H.P. 39,000. Speed 36 knots. Oil fuel 505 tons.

Progress was very slow and it was not until 11.4.32, that a single engine-room design based on CODRINGTON was accepted by the Board. The two-engine-room design was regarded as being less handy, less flexible and more expensive — it would cost more than £15,000 extra to build than the single-engine-room design. However, the two-engine-room design had the benefit that one hit would not put the ship out of action.

Tenders for the vessel were invited on 4.7.32 and received 15.8.32. EXMOUTH, as the vessel was now named, was ordered from Portsmouth Dockyard on 1.11.32. Her machinery was ordered the same day from Fairfield's at a cost of £156,235.

Her dimensions proved to be very close to Design A.

Detail	EXMOUTH	DUNCAN	CODRINGTON
Length (OA)	338'	326'	340'
Beam	33½'	33'	33¾'
Mean Draft	8¾'	8$\frac{7}{12}$'	8$\frac{15}{16}$'
Standard Displacement	1,495 tons	1,400 tons	1,520 tons
S.H.P.	38,000	36,000	39,000
Deep Speed	32 knots	32 knots	31½ knots
Oil Fuel	480 tons	470 tons	430 tons
Endurance	5,700 miles	5,600 miles	5,000 miles

EXMOUTH and DUNCAN had the same armament of four 4.7", one 3" high angle, two 0.5" machine guns, four Lewis guns, eight 21" torpedo tubes and 20 depth charges. CODRINGTON differed in having five 4.7", with the 3" high angle and 0.5" machine guns being deleted.

CATEGORY (Weights in tons)	EXMOUTH	DUNCAN	CODRINGTON
General Equipment	116	110	89
Armament	131	145	133
Machinery	550	575	575
Hull	698	690	688
Standard Displacement	1495	1520	1485

There was no Board Margin. On 6.10.34, when inclined, EXMOUTH's displacement was 1,393 tons and her estimated deep displacement was 2,053 tons.

THE DESTROYERS: THE ECLIPSES

At the Conference on 9.4.31, the following parameters of the proposed vessels were agreed:
(i) To be repeats of the C/D design.
(ii) Two vessels to be fitted for minelaying whilst the others would be fitted with ASDIC and T.S.D.S.
(iii) The engine room to be one frame space longer than that in the C/D's, but the boiler-room space to be one frame less.
(iv) It was hoped to make No. 2 gun a high angle weapon. (A final decision could not be made until 12.31.)
(v) The boilers and machinery to be the same weight.
(vi) Storage space for 12 weeks provisions to be provided.
(vii) Bullet-proof plating to be provided, as in the C/D's.

A sketch design and Legend of Particulars was submitted on 18.6.31 in line with the Controller's proposals. Oil storage of 445 tons was provided to increase endurance to 5,500 mile at 15 knots — the same as in the CRUSADERs but 700 miles more than the ACASTAs. Allowance was made in the magazines for the stowage of an extra 100 rounds of 4.7" high angle ammunition, in the event of the 60° 4.7" being fitted. On 11.7.31, the three-boiler design was adopted.

	T.S.D.S. '31 DESTROYER	ACASTA as built	CRUSADER as designed	1931 Destroyer at 3.2.32	
				With T.S.D.S.	As Minelayer
Length (P.P.)	310'	312'	317¾'	318¼'	318¼'
Length (W.L.)	318'	320'	326'	326'	326'
Breadth (Ext.)	32½'	32¼'	33'	33¼'	33¼'
Mean Draft (Std.)	8$\frac{7}{12}$'	8½'	8½'	8$\frac{7}{12}$'	8⅔'
Displacement (tons)	1,370	1,350	1,375	1,405	1,415
S.H.P.	35,000	34,000	36,000	36,000	36,000
Speed (deep)	32 knots	32 knots	32 knots	31½ knots	31½ knots
Oil Fuel (tons)	445	390	470	470	470
Armament					
4.7"	4	4	4	4	2
2 pounder pom-pom	—	2	—	—	—
3" H.A.	1	—	1	1	1
0.5" machine guns	2	—	2	2	2
Lewis Guns	4	4	4	4	4
21" T. Tubes	8	8	8	8	—
Mines	—	—	—	—	60 Type XIV

Comparative Weights (All tons)	1931 DESTROYER WITH T.S.D.S.	1931 DESTROYER AS MINELAYER
General Equipment	87	86
Armament	128	131
Machinery	525	525
Hull	665	673
Standard Displacement	1,405	1,415

As late as 11.31 a single funnel arrangement was under consideration, but as with the leader, was not proceeded with. On 4.12.31, it was decided that every destroyer was to be fitted with ASDIC, instead of alternative flotillas as had been previous practice. It was realised that anti-submarine operations would be much easier if all destroyers were to be fitted with ASDIC. Furthermore: "with the smaller number of destroyers which we shall possess under the provisions of the London Naval Treaty, it seems the more necessary that all should be available for A/S operations." It was estimated that the cost of providing an installation for each vessel would be £2,500.

The design was approved by Board Minute 2921 of 18.2.32 and 11 days later a decision was made to fit one minelaying and one T.S.D.S. destroyer as divisional leaders. Thus, the leader and six destroyers were to be fitted with ASDIC and T.S.D.S. Subsequently, it was agreed on 17.6.32 that destroyers of the 1932, 1933 and 1934 programmes were to be fitted with T.S.D.S. and ASDIC.

On 22.6.32 the Controller issued a verbal instruction that arrangements were to be made in the tender drawings and specification of the 1931 destroyers to provide for 4.7" guns with 40° elevation, as the provision for the 60° guns had been dropped.

Tenders were sought on 4.7.32 and were received by 15.8.32. Orders were finally placed on 1.11.32. Details still exist on the contract prices of six of the destroyers.

Vessel	Builder	Total Cost	Hull Cost	Main Machinery	Aux. Machinery
ESCAPADE	SCOTTS	£249,987	£105,510	£136,820	£7,657
ESCORT	SCOTTS	£249,587	£105,110	£136,820	£7,657
ELECTRA	HAWTHORN LESLIE	£253,350	£105,470	£139,916	£7,964
ENCOUNTER	HAWTHORN LESLIE	£252,250	£105,370	£138,916	£7,964
ECHO	DENNY	£247,009	£102,965	£135,465	£8,579
ECLIPSE	DENNY	£246,664	£102,620	£135,465	£8,579

THE MINELAYING DESTROYERS

Cover 504 gives details of the discussions that took place over the design of the last two vessels of the programme that were to be fitted for conversion to minelayers. In a memo of 14.7.31, the D.N.C. outlined the details of the destroyers that embodied the decisions of the Controller's Conference of 9.4.31. The vessels were to have the same characteristics as the other 1931 programme destroyers, except "when engaged as a minelayer, with a full complement of mines aboard the following items would *not* be on board"
(i) The forward torpedo tubes and torpedoes.
(ii) No. 1 and No. 4 4.7" guns.
(iii) Shell and ammunition for No. 1 and 4 guns.
(iv) Twin-speed destroyer sweep, winches and equipment.
(v) Starboard whaler and davits.
(vi) The torpedo store would become the mining store.

The mining equipment would consist of:
(i) Two sets of wide gauge fixed rails to carry 60 Mk XIV mine units.
(ii) Installation of mechanical chain conveyor laying gear.
(iii) The after deck house would be modified to provide adequate fore and aft gangways and overhead clearances when mines were carried.

A trial of the minelaying equipment had taken place on the destroyer SKATE on 9.6.31. The minelaying equipment was to be of the chain-conveyor type fitted in the submarine PORPOISE and for trial purposes in the cruiser ADVENTURE.

The legend and drawings of these destroyers of the 1931 programme were also approved by Board Minute No. 2921 of 18.2.32.

On 29.4.32 the vessels' anti-aircraft armament was amended, with the omission of the 3" high angle gun and the improvement of the arcs of fire of the 0.5" which were to be mounted on separate platforms amidships, followed by the removal of the 27' whaler and the 23' cutter to the forecastle as with the leader. On 30.5.32 it was confirmed that when the high angle 4.7" mounting was available (it never appeared in service) it would be mounted in No. 3 gun position.

Tenders for the two vessels were invited on 4.7.32, the tender of Swan Hunter and Wigham Richardson dated 15.8.32 was accepted and the vessels ordered on 1.11.32. Tender prices were as follows:

Name	Total Cost	Hull	Main Machinery	Auxiliary Machinery
ESK	£247,649	£106,225	£133,500	£7,924
EXPRESS	£247,279	£105,855	£133,500	£7,924

The armament statement was finally agreed during 8.32 as 4 4.7" (Mk. IX), two machine guns, four Lewis guns, 0.3" Vickers machine guns, 60 mines and 20 depth charges.

On 1.11.34 ESK laid 60 mines off Spithead "without delay or difficulty" but loading took $3\frac{1}{2}$ hours.

BUILDING PARTICULARS: 1931 PROGRAMME

Name	Builder	Ordered	Laid Down	Launched	Commissioned
ECLIPSE	DENNY	1.11.32	22.3.33	12.4.34	29.11.34
ECHO		1.11.32	20.3.33	16.2.34	22.10.34
ENCOUNTER	HAWTHORN LESLIE	1.11.32	15.3.33	29.3.34	2.11.34
ELECTRA		1.11.32	15.3.33	15.2.34	13.9.34
ESCORT	SCOTTS	1.11.32	30.3.33	29.3.34	30.10.34
ESCAPADE		1.11.32	30.3.33	30.1.34	30.8.34
EXPRESS	SWAN HUNTER & WIGHAM RICHARDSON (Wallsend Slipway)	1.11.32	24.3.33	29.5.34	2.11.34
ESK		1.11.32	24.3.33	19.3.34	28.9.34
EXMOUTH	PORTSMOUTH DOCKYARD (Fairfield)	1.11.32	15.5.33	30.1.34	9.11.34

EXMOUTH at the time of her loss, by torpedo from U22 in the Moray Firth on 21.1.40, was little altered from this pre-war photograph.

EXMOUTH (H02)

EXMOUTH commissioned on 12.11.34 as leader of the 5th D.F. of the Home Fleet. She was to remain with this flotilla until the re-organisation of the destroyer Flotillas in 4.39. The 5th D.F. was attached to the Mediterranean Fleet for the Abyssinian crisis between 8.35-3.36, when she returned to the U.K. The outbreak of the Spanish Civil War meant that the Home Fleet Flotillas were soon engaged on non-intervention patrol duties in Spanish waters. EXMOUTH participated in these duties between 9-11.36 off the Biscay ports, 1-3.37 off the Mediterranean coast and 10.37-1.38 again off the Mediterranean ports. During this period, EXMOUTH refitted at Alexandria between 4.10.35-5.1.36 and at Portsmouth between 17.11.36-19.1.37 and again between 21.11.38-16.1.39. This latter refit was followed by three months of exercises off Gibraltar.

On 28.4.39, EXMOUTH reduced to a special complement, but remained under the control of the Rear Admiral Destroyers. The next few months were spent as a boys' training ship and on aircraft co-operation duties with Portsmouth Command.

On 2.8.39 she was brought forward into full commission for service with the Reserve Fleet and on the outbreak of war EXMOUTH was leader of the 12th D.F., initially with the Home Fleet. In 12.39 she joined the Western Approaches Command and a month later the Rosyth Command. She was only to survive a few more days, as on 21.1.40 she was torpedoed by U22 in the Moray Firth and was lost with all hands (her C.O. Capt. Benson, 14 officers and 174 ratings). She lies in position 58° 15' N 02° 06' W.

ECHO/R.H.N NAVARINON (H23)

On commissioning on 25.10.34, she joined the Home Fleet's 5th D.F. and immediately participated in that Fleet's cruise of the West Indies between 1-3.35. This marked the end of "normal" peace duties as during 9.35 the whole of the 5th Flotilla, ECHO included, were attached to the Mediterranean Fleet for the Abyssinian crisis that ended in 3.36. However, on 19.11.35 ECHO and her sister ENCOUNTER were in collision during exercises off Alexandria. Repairs to ECHO's badly buckled bow were not completed at Malta until 19.1.36.

She returned to the U.K. and operations with the Home Fleet until 1.37, when she became part of the non-intervention patrol off the Spanish Biscay ports until 18.3.37. She then returned to her normal duties and ECHO refitted at Devonport between 9.8-7.10.37 before operating with the French Fleet at Oran during the last month of that year.

Further service in Home waters then followed, which was punctuated by a fire in boiler room 2 on 31.5.38. Repairs to her damaged ring-mains and electrical leads were completed at Portsmouth 18.6.38. A further refit followed at Devonport between 2.11.38-11.1.39, before ECHO saw her last period of service in Spanish waters, being based at Gibraltar until 10.3.39. ECHO and ESK were in collision in night exercises off Gibraltar on 6.2.39, with ECHO receiving slight damage.

ECHO resumed her duties in Home waters for the last six months of peace and was scheduled to be replaced by JAGUAR in the Flotilla, which was renumbered the 7th in 4.39. Still a member of the 7th Flotilla on 3.9.39, she was lying at Sheerness for boiler repairs when the war broke out. After being replaced by JAGUAR, she transferred to the 12th D.F. of the Western Approaches Command until 6.40. She had the task of rescuing survivors from the aircraft carrier COURAGEOUS torpedoed by U29 off western Ireland on 17.9.39.

On 13.10.39, ECHO grounded whilst entering Plymouth in fog. Damage repairs were not completed until 12.11.39. She then undertook normal duties until 20.2.40 when she started a refit at Leith, which was not completed until 16.4.40. She was then loaned to the Home Fleet on 1.5.40 until mid 6.40 for escort duties based at Scapa. She assisted the torpedoed A.M.C. SCOTSTOUN on 13.6.40 in position 57°00'N 09°57'W and later rescued survivors from the ARANDORA STAR torpedoed by U47 in position 56°20'N 10°40'W.

During 6.40, ECHO was transferred to the 3rd D.F. of the Home Fleet until 8.41. She escorted the troop transport ULSTER PRINCE to Iceland and back between 9-19.7.40 and then operated on escort duties around Scotland until 28.8.40. On that day, she left Scapa as one of the escorts for the vessels for Operation 'MENACE' — the ill-starred Dakar expedition. She did not return to the Clyde until 27.10.40, when she arrived with survivors from the liner EMPRESS OF BRITAIN, sunk 70 miles north-west of Donegal Bay by submarine and air attack.

ECHO reported major defects on her arrival and extensive machinery repairs were undertaken at Barclay Curle's yard on the Clyde between 30.10-15.12.40. She then returned to the Home Fleet at Scapa and immediately participated in Operation 'RUBBLE' — the escape of six Norwegian vessels from Gothenburg to the U.K. in 1.41. She acted as one of the escorts for the raiding force to the Lofoten Islands between 1-16.3.41. Three days after her return, she picked up 38 survivors from the Norwegian vessel LEO. On 13.4.41, she was sent to assist the A.M.C. RAJPUTANA, torpedoed by U168 in position 64°50'N 27°25'W.

On 21.5.41, ECHO was despatched to escort British cruisers on patrol in the Denmark Strait, hunting the BISMARCK, but had to return to Iceland to refuel three days later. On 25.5.41 she left Iceland with ELECTRA to escort the battleship PRINCE OF WALES, damaged in action, back into Hvalfjord.

ECHO refitted at Harland and Wolff's North Woolwich yard between 14.8–4.11.41 before returning to service with the 3rd D.F. of the Home Fleet until 5.42. She immediately began operations in Icelandic waters and escorted convoys PQ6 and QP4 to and from Murmansk between 13.12.41 and 11.1.42, when she returned to Scapa. Three months of escort duties between Iceland, Scapa and Rosyth then followed, before she left the Clyde on 14.4.42 as as convoy escort to Gibraltar. At Gibraltar she joined the U.S. aircraft carrier WASP on a flying-off operation to Malta between 19-21.4.42. A second flying-off operation with the aircraft carrier EAGLE took place between 7-10.5.42, before ECHO returned to the U.K. for further Icelandic convoy duties until 18.6.42.

ECHO in this photograph dated 30.8.42 shows only the initial war modifications with the after bank of torpedo tubes replaced by a 3" H.A. gun and the addition of Oerlikons.

On this date she started a refit at Hull which was not completed until 31.8.42. She then joined the 8th D.F. of the Home Fleet. She immediately participated in convoy operations PQ18 and QP14 and during 11.42 she escorted QP15 from North Russia. Her escort duties in northern waters continued until 2.43.

She refitted on the Humber between 2.43 and 6.43 and on 17.6.43 she left Scapa with other members of the 8th D.F. for service with the Mediterranean Fleet until 2.44. She participated in the invasion of Sicily on 10.7.43 and three days later, with the destroyer ILEX, she sank the Italian submarine NEREIDE in the Straits of Messina. Two months later she participated in Operation "AVALANCHE" — the invasion of Italy at Salerno. She then operated on escort duties in the Eastern Mediterranean for the remainder of the year. The critical situation in the Aegean meant that ECHO conveyed personnel and stores to Leros and other Aegean Islands.

On 13.11.43 ECHO and BELVOIR rescued 109 survivors from the destroyer DULVERTON which had been sunk by glider bombs off Kos. The next day ECHO and BELVOIR bombarded positions on Mount Appetici on Leros, after the island had been invaded. On the night of 14-15.11.43, the two destroyers collected troops from Samos and ECHO landed more troops at Leros.

The next month, she operated off the west coast of Italy — in the Gulf of Gaeta in support of ground forces and participated in a feint landing in that place on the night of 12/13.12.43. On 22.12.43, she berthed at Malta and started a refit, which was not completed until 4.44. ECHO formally paid off on 5.4.44 and was transferred to the Greek Navy as NAVARINON.

ECHO is seen as the Greek NAVARINON at the Coronation Review during 6.53. She is little altered from wartime.
(National Maritime Museum N32362)

She operated on convoy escort and bombardment duties in the eastern Mediterranean, being based at Piraeus from 11.44 until the end of the war. She was also active in supporting the pro-Western forces against the Communist insurgents at the start of the Greek Civil War. She was, therefore, retained in the Greek Navy post war on these and later on training duties. She represented Greece at the Coronation Review in 1953 and was not returned to the Royal Navy until 8.3.56. at Malta. She was towed to the UK and arrived at the yard of Clayton and Davie at Dunston on Tyne on 26.4.56 for demolition. BISCO confirmed on 9.4.57 that demolition had been completed.

ECLIPSE (H08)

ECLIPSE commissioned on 1.12.34 for service with the 5th D.F. of the Home Fleet and operated in home waters until she left Plymouth on 31.8.35 to reinforce the Mediterranean Fleet at the start of the Abyssinian crisis. She served in the eastern Mediterranean until 3.36, being stationed at Haifa between 8.9-3.10.35, then Alexandria between 4.10.35-5.1.36 and then at various locations in the eastern Mediterranean. On her return to the UK, she refitted at Devonport between 20.3-30.4.36.

She then spent the next three years with 5th D.F. mainly in Home waters, before reducing to Reserve at Devonport on 5.5.39 as tender to the cruiser COLOMBO. During this period, she served off the Bay of Biscay ports between 1-3.37, made a tour of Scandinavia between 18.6-6.7.37, undertook patrol duties based on Oran and Gibraltar between 10.37-12.37, followed by further patrol duties on the southern coast of Spain between 1-3.38.

ECLIPSE remained in Reserve until 2.8.39, when she recommissioned for the Review of the Reserve Fleet. On the outbreak of war ECLIPSE joined the 12th D.F. and by 12.39 was serving with the Western Approaches Command. Previous to this, ECLIPSE with ECHO and EXMOUTH escorted the battle-cruiser HOOD, which was attached to a French Squadron, on a sortie south of Iceland attempting to locate the German battle-cruisers SCHARNHORST and GNEISENAU then raiding in the Atlantic. ECLIPSE continued to undertake escort duties and on 8.4.40 whilst escorting convoy ON25, she was ordered to join the Home Fleet.

She was only to be on these duties three days, as she was severely damaged by air attack whilst acting as part of the anti-submarine screen for the fleet off Norway. Her engine-room was flooded after a bomb had landed between stations 104-112. Gland spaces, shaft tunnels and warhead magazine were also flooded. She was immobilised and had to be towed to Lerwick. ECLIPSE then repaired on the Clyde until 10.8.40.

She immediately joined the Home Fleet's 3rd D.F. for the next 22 months. ECLIPSE, with INGLEFIELD, ECHO and ESCAPADE left the Clyde as part of the escort of the convoy for Operation 'MENACE', but had to leave prematurely and returned to Sierra Leone to repair defects that had arisen. The repairs continued until 14.10.40, when she left for Gibraltar where further machinery repairs were undertaken between 28.10-15.11.40. On completion of these repairs ECLIPSE returned to the UK and her duties with the 3rd D.F. She refitted at Devonport between 12.4-4.6.41 and again at Hull between 6.12.41-21.2.42.

A fine view of ECLIPSE, shortly after first commissioning in 1935. She was to be mined in the Aegean on 24.10.43 with heavy loss of life.

ECLIPSE was then immediately embroiled in Arctic convoy duties. On 29.3.42 she was escorting the seven ships of the convoy PQ13 and was on station 1,500 yards ahead of the convoy. At 09.50 hours a ship was sighted and recognised as German and a running action ensued. 5 minutes later a large explosion occurred on the enemy vessel. At 10.20 hours the enemy ship was sighted stopped in position 72°98'N 30°09'E with her stern awash and considerable damage aft and amidships. ECLIPSE was about to fire her torpedoes at the vessel, when 2 enemy destroyers were sighted at a range of two miles and fire was opened. At 10.28 hours, she was hit twice — one shell detonated on deck and the second entered the after superstructure and detonated 4.7" charges and caused serious damage and heavy casualties to the ammunition supply party. At 10.31 hours 'A' gun was put out of action and a hole made in the starboard side. Two ratings died of wounds and 7 others were wounded.

Action damage repairs in North Russia were completed on 6.4.42 and she returned to the UK for further repairs. These were completed at Devonport by 6.5.42. Repairs included plating the ship's side between stations 13-28, repairs to the oil fuel tanks, shrapnel damage to the upper and lower decks as well as the repair of two of her 4.7" guns.

ECLIPSE then joined the Home Fleet's 8th D.F. during 7.42 and was to remain with this flotilla until her loss. She participated in the escort of convoy PQ18 to Russia in 9.42, refitted on the Humber between 27.9-20.11.42 and was loaned to the Western Approaches Command during 4-5.43 at the height of the U-boat crisis.

On 28.6.43 she arrived at Gibraltar as part of the reinforcements for the Mediterranean Fleet for the Sicily invasion. She was based at Malta, but later transferred to Alexandria a few days before her loss at the start of the Aegean campaign. On 24.10.43 ECLIPSE was mined in position 37°01'N 27°11'E. A violent explosion occurred abreast of the boiler room under the ship on the starboard side. She quickly listed to port until she lay on her beam-ends and sank in about three minutes. She seemed to break in two abreast the bridge. It seems probable that there were two explosions with one or two seconds between, the second being a large magazine explosion. Five officers and 114 ratings were killed.

A pre-war view of ELECTRA. (W.S.P.L.)

ELECTRA (H27)

After commissioning, ELECTRA served with the 5th D.F. of the Home Fleet until 11.38. However, on 18.7.36 she was damaged whilst entering dock at Sheerness and was under repair there until 5.9.36. She then spent 9-11.36, 1-3.37 and 10-12.37 on non-intervention patrol duties in Spanish waters. On 17.11.38 she started a refit, but major problems with her turbines and superheaters meant that repairs were not completed until 18.3.39. She then entered Reserve at Chatham until 2.8.39, when she commissioned for the Reserve Fleet Review.

On the outbreak of war, ELECTRA was allocated to the 12th D.F. at Portland, undertaking escort duties in the Western Approaches between 10.39-4.40 before she became embroiled in the Norwegian Campaign for the next two months. ELECTRA and ESCORT picked up the survivors of the Donaldson liner ATHENIA, that had been sunk by U30 (Lemp) within hours of the outbreak of war.

On 13.6.40, ELECTRA collided with ANTELOPE off Trondheim and repairs and modifications at Troon were not completed until 31.8.40. She then transferred to the 3rd D.F. of the Home Fleet, escorting with the destroyer JACKAL the minelaying operations SN41 on 11.9.40. Some two months later, on 6.11.40, ELECTRA escorted the battle-cruiser REPULSE, searching for the raider ADMIRAL SCHEER after the latter had sunk the A.M.C. JERVIS BAY protecting her convoy HX84 from total destruction.

ELECTRA was to spend the next year on escort duties with the Home Fleet and on Arctic convoy duty. The most important actions during this period were her participation in the convoying of six Norwegian Merchant vessels that had escaped from Sweden (Operation "RUBBLE") on 25.1.41, and the search for the German battle-cruisers SCHARNHORST and GNEISENAU between 26-30.1.41. She participated in an anti-raider sweep between 8-10.2.41 with the QUEEN ELIZABETH and HOOD between 19-23.3.41.

ELECTRA participated in the BISMARCK operation between 23-27.5.41 and was present when the HOOD was sunk on the 24.5.41. After BISMARCK had been sunk, ELECTRA was refitted in the Royal Docks at London by Green & Silley Weir between 24.6-5.8.41. ELECTRA then operated in Arctic waters between 8-10.41, escorting convoys to North Russia.

After minor repairs at Rosyth, ELECTRA and EXPRESS left Scapa on 23.10.41 as part of the escort for the battleship PRINCE OF WALES and the battle-cruiser REPULSE for their passage to Singapore, where they arrived on 2.12.41. Eight days later ELECTRA had the task of rescuing the survivors from the two capital ships, after they had been attacked and sunk by Japanese aircraft off Kuantan, eastern Malaya.

After returning the survivors to Singapore, ELECTRA operated in the Singapore area on convoy escort duties, being attacked, without damage, by enemy aircraft in the Banka Strait on 28.1.42.

On 26.2.42, ELECTRA with ENCOUNTER and JUPITER were escorting the cruiser EXETER on convoy duty off Batavia when the squadron was ordered to join Admiral Karel Doorman's battle squadron at Sourabaya. After an unsuccessful sortie on the night of 26/27.2.42, the allied fleet returned to Sourabaya. When new reports of Japanese movements were received ELECTRA, ENCOUNTER and JUPITER were sent ahead of the fleet. At 16.00 hours on 27.2.42 the action opened and after EXETER had been damaged and the Dutch destroyer KORTENAER sunk, the fleet withdrew. In poor visibility caused by smoke, the opposing destroyers became involved in engagements at short range during which ELECTRA was sunk by gunfire from the Japanese cruiser JINTSU and destroyers, with the loss of six officers and 102 ratings. Seven other ratings were to die in captivity from the 54 originally saved from the ship.

ENCOUNTER (H10)

ENCOUNTER lived up to her name, as she suffered no less than three collisions during her pre-war service! After working up, she joined the 5th D.F. of the Home Fleet during 12.34 and immediately participated in a deployment to the West Indies between 1-3.35. ENCOUNTER and ESCAPADE collided off Portland on 18.6.35 and ENCOUNTER was under repair at Devonport between 19.6-8.7.35.

Two months later she was one of the reinforcements sent to the Mediterranean Fleet for the duration of the Abyssinian crisis. However, on 19.11.35 during night exercises off Alexandria ENCOUNTER collided with ECHO. ENCOUNTER was slightly holed in No. 2 boiler room and returned to Alexandria. Collision repairs were completed at Malta between 29.11.35-8.2.36. The 5th D.F. returned to the U.K. the next month and remained in U.K. waters until the Flotilla participated in the non-intervention patrol duties in the Bay of Biscay during 1-3.37.

ENCOUNTER then participated in the Home Fleet Scandinavian tour, between 21.6-6.7.37, before refitting at Sheerness between 30.7-18.11.37. After working up at Portland she operated along the south coast of

ENCOUNTER at Torquay on 5.6.38. She sports the distinctive tricolour on No. 2 Mounting, used for recognition purposes off the coast of Spain. (National Maritime Museum N9481)

Spain during 2-3.38, before returning to Home waters and being extensively damaged in a collision on 26.9.38. Her bow from No. 9 bulkhead to her stem was destroyed and repairs were undertaken at Hawthorn Leslie's Hebburn yard, between 27.9-10.38. After repairs, she was based at Gibraltar for Spanish service between 1-3.39.

On 15.7.39, ENCOUNTER was transferred to the C.in C. Nore at the start of a refit and was then to have entered Reserve. However, the imminence of war resulted in her commissioning as part of the 12th D.F. with a crew consisting of 60% of reservists.

ENCOUNTER spent the first nine months of the war on escort duties with the Western Approaches Command. She was based at Milford Haven until 10.39, then Plymouth for the next two months, before being based at Scapa Flow. ENCOUNTER was then embroiled in the Norwegian Campaign. A special duty was to provide A/S protection with MASHONA to the First Minesweeping Flotilla, which was collecting buoys from the wreck of the boom-carrier ASTRONOMER lost off Kinnaird Head on 2.6.40.

ENCOUNTER refitted at Sheerness between 20.6-20.7.40, which included de-gaussing. On that day, she sailed for Gibraltar as escort for the troopships REINA DEL PACIFICO and CLAN FERGUSON and the aircraft carrier ARGUS, and there joined the recently formed Force 'H'. ENCOUNTER was based at Gibraltar as a unit of the 13th D.F. until 12.40 and during this period participated in the Cape Spartivento action.

After escorting the aircraft carrier FURIOUS to Freetown and Takoradi during 1.41, she spent a period in the South Atlantic before she made passage to join the Mediterranean Fleet and arrived at Alexandria on 12.4.41. This period of service was to last less than three weeks. Whilst on blocks in No. 2 Dock at Malta, she was hit by bombs on 29.4.41, 30.4.41 and 16.5.41. The first bomb exploded on the dock steps and caused much blast and splinter damage. The second bomb, which hit at 01.30 hours on 30.4 penetrated the forecastle, upper and lower decks and exploded in the A/S directing compartment and blew a hole in the hull.

The third hit at (04.30 hours on 16.5.41) penetrated the 0.5" machine gun platform and the upper deck and exploded in the port bilges of No. 2 boiler room, blowing another hole in the hull. There were no casualties as the ship was evacuated at night. The result of these hits was that the vessel was unseaworthy, with her 4.7", 3", 0.5" guns and torpedo tubes out of action. The Nos. 1, 2 and 3 boilers and cruising turbines were out of action with the ship flooded between 9 and 41 bulkheads. Repairs to her boilers took until 7.41, when on the 26th of that month, she arrived at Gibraltar under the cover of Force 'H'. She then remained at Gibraltar, but again transferred to the South Atlantic Command during 9.41 for a month.

On 16.10.41 she returned to Alexandria and operated with the Mediterranean Fleet for a month, before being transferred to the Eastern Fleet. She left Alexandria for Singapore on 14.11.41, arriving at Singapore on 3.12.41. She operated from there until 20.1.42, when she joined the 7th D.F. in the Dutch East Indies.

ENCOUNTER participated in the Java Sea action, where the cruiser EXETER was severely damaged on 27.2.42. On 1.3.42, the EXETER left Sourabaya for Colombo with the ENCOUNTER and the American destroyer POPE as escorts. Spotted by Japanese reconnaissance aircraft, the vessels were engaged by four Japanese cruisers and three destroyers. EXETER ordered the destroyers to disengage, but ENCOUNTER was unable to do so because her engine room had had to be abandoned due to a fire caused by a broken oil pipe and with all power lost her guns could not train. The vessel was scuttled by her crew and sank in a position 04°30' S 111°00' E with the loss of one officer and seven ratings. The majority of her crew became prisoners of war, with 37 ratings dying in captivity.

ESCAPADE (H17)

On commissioning, ESCAPADE joined the 5th D.F. of the Home Fleet, until 6.39. However, she was to see several periods of service overseas participating in a cruise to the Caribbean between 1-3.35 on attachment to the Mediterranean Fleet for the Abyssinian crisis between 9.35 and 3.36 and no less than five detachments to Spanish waters between 1-3.37, 7.37, 10-12.37, 1-3.38 and 2-3.39.

On 18.1.39 ESCAPADE collided with her sister ECLIPSE. A breakdown in her engine room resulted in an immediate loss of speed and ECLIPSE stationed astern had no chance to stop and collided with ESCAPADE at frame number 167 on the starboard side. Repairs took 17 days.

On 16.6.39 ESCAPADE reduced to Reserve as tender for the cruiser COLOMBO at Plymouth. She did not remain there long as on 2.8.39 she was brought forward to full commission for the Reserve Fleet Review three days later.

On the outbreak of war, ESCAPADE was a member of the 12th D.F. and served on escort and patrol duties until 4.40. On 5.12.39 she picked up survivors of the S.S. NOVA SCOTIA; unsuccessfully attacking two U-boats on 5.11.39 and 15.11.39 in the Channel. However, on 25.2.40 whilst escorting convoy HN14 for Norway, ESCAPADE forced U63 to the surface, where the latter scuttled herself. ESCAPADE picked up survivors. She then operated in Norwegian waters between 7.4.40 and 17.6.40. On that day, she started to escort the aircraft carrier ARK ROYAL to Gibraltar to join the newly-formed Force 'H' and remained at Gibraltar until 8.40 when she was ordered home. Although she had returned to home waters to undertake anti-invasion duties with 3rd D.F., she remained in U.K. waters only three weeks, before being detached to escort the battleship BARHAM during the unsuccessful Dakar operation.

On her return to the U.K., ESCAPADE spent the next eight months on escort duties and in 1.41 she participated as an escort during Operation 'RUBBLE'. She also escorted the battleship NELSON in an unsuccessful search for the German battle-cruisers SCHARNHORST and GNEISENAU, that were raiding in the North Atlantic during 2.41. On completion of her escort duties, with HX125, she refitted on the Tyne between 26.5-10.7.41.

On completion of this refit, ESCAPADE rejoined the Home Fleet's 4th D.F. and immediately provided part of the escort for the battleship PRINCE OF WALES when the latter carried Mr. Churchill to his meeting with President Roosevelt at Argentia, Newfoundland on 10.8.41. ESCAPADE then escorted several Arctic convoys, arriving at Archangel with the first convoy on 31.8.41, and remained on these duties until she refitted at Immingham from 9.2-30.3.42.

In 4.42, ESCAPADE escorted the Norwegian tanker LIND into Methil on her escape from Sweden. 5.42 was to see ESCAPADE covering the unsuccessful attempt to get the damaged cruiser TRINIDAD to Iceland from North Russia. ESCAPADE then formed part of the escort for QP12 from Kola, which had an uneventful passage to Iceland.

On 5.6.42, she was transferred to the 5th D.F. and immediately escorted Convoy WS19Z to Malta, as part of Operation 'HARPOON'. After escorting the aircraft carrier ARGUS to the U.K., ESCAPADE formed part of the escort for the covering force for the ill-fated convoy PQ17 to Russia. After this hectic activity, she refitted at Liverpool between 20.7-24.9.42.

ESCAPADE then escorted the carrier FURIOUS to Gibraltar, arriving on 25.10.42, before escorting 'TORCH' convoys KMF1, and KMF2. She returned to the U.K. with MKF1(X) and arrived at Greenock on 19.11.42. After repairing on the Thames between 27.11-23.12.42, ESCAPADE joined Escort Group B3 in the Western Approaches Command. During the next six months, she escorted convoys in the Atlantic, notably convoys HX228, ONS175, HX232 and HX239 in May 1943.

A refit at Cardiff, including the fitting of Hedgehog, was completed between 3.6-5.9.43. After a quick work-up, she joined convoy ONS18 and after gaining a submarine contact off Northern Ireland on 20.9.43, ESCAPADE fired her Hedgehog at the submarine. An explosion occurred, with the Hedgehog mounting in 'B' gun position being wrecked and the bridge and wheelhouse badly damaged. The explosion was caused by a premature explosion of a projectile after firing which countermined those remaining on the mount. Casualties were heavy, with 15 ratings killed, one dying of wounds and nine others wounded. Repairs at Portsmouth were not completed until 30.12.44.

ESCAPADE on 12.2.45 after the completion of extensive repairs following the explosion on her Hedgehog mounting on 20.9.43. She is fitted with a Squid mounting in lieu of A gun and carries, unusually, a Type 277 radar.

ESCAPADE then joined the 8th Escort Group for the last months of the war and was fitted with the Squid system for trials purposes. After escorting Norwegian personnel back to Norway during 5.45, she joined the A/S Training Flotilla. ESCAPADE was approved for scrap on 18.2.46. On 15.11.46 her equipment was ordered to be removed. On 3.12.46 she paid off as tender to H.M.S. TARTAR. She was handed over to BISCO on 17.5.47 for scrapping and broken up at the Grangemouth yard of G. W. Brunton from 3.8.47.

ESCORT (H66)

On commissioning on 6.11.34, ESCORT joined her sisters in the 5th D.F. of the Home Fleet until 3.39. A tour of the West Indies between 1-3.35, was followed by a refit at Sheerness between 27.3.35 and 30.4.35. Between 9.35 and 3.36 ESCORT was attached to the Mediterranean Fleet during the Abyssinian crisis. However, after hitting a lock at Sheerness, ESCORT was under repair for seven weeks until 5.9.36.

The years 1936 to 1939 saw ESCORT spending large periods of time in Spanish waters with the Non-Intervention patrol, the periods 9-11.36, 6.37, 10-11.37 being spent off the Biscay ports, whilst 1-3.37, 11-12.37, 1-3.38 and 1-3.39 were spent around Gibraltar and the Spanish Mediterranean coast.

During the Munich crisis of 9.38, ESCORT was based at Scapa and Invergordon with the Home Fleet. After refitting at Sheerness until 1.39, ESCORT spent her final period of service in Spanish Waters, based at Gibraltar. Returning to the U.K. on 24.3.39, ESCORT commissioned as a tender to the cruiser CALEDON in the Reserve Fleet on 5.5.39. However, on 2.8.39, she was brought forward into full commission with the 12th D.F. in which she was to serve until her transfer to the Mediterranean Fleet.

ESCORT was the first British destroyer lost in the Mediterranean, when torpedoed by the Italian submarine MARCONI on 8.7.40, and finally sank three days later.

ESCORT with her sister ELECTRA, only hours after the declaration of hostilities, rescued over 300 survivors from the torpedoed Donaldson liner ATHENIA in the North Channel. ESCORT spent the next eight months on normal destroyer duties, firstly in the Western Approaches Command and then based at Rosyth from 12.39. She had refitted at Falmouth between 10.1-12.2.40 and continued her duties until 11.5.40 when she was slightly damaged in collision with the Polish liner CHOBRY.

In 5.40 ESCORT was based at Scapa to escort vessels of the Home Fleet and these duties continued until she left for Mediterranean on 26.6.40, arriving at Gibraltar on 2.7.40.

Her career with the Mediterranean Fleet was to be all too brief as at 02.15 hours on 8.7.40 she was torpedoed by the Italian submarine MARCONI, north of Cyprus. The torpedo struck ESCORT on her starboard side between No. 1 and No. 2 boiler-rooms, causing a 20ft hole from 4ft below the upper deck to below the bilge keel. All compartments forward of No. 41 bulkhead flooded immediately, but the vessel was towed for three days. However, at 06.30 hours on 11.7.40 she suddenly developed a 30° list to port and was abandoned. The vessel turned over on her beam-ends, broke her back between No. 1 and No. 2 boiler rooms and finally sank at 11.14 hours that day. Luckily only two of her complement were killed.

ESK was one of a pair of 'E' class destroyers completed as mine-layers. Number 1 and 4 guns have been landed, in anticipation of stowing mines.

ESK (H15)

ESK on commissioning on 2.10.34 joined her sisters in the 5th D.F. of the Home Fleet until 6.39. However, with the increase in tension apparent in Europe after 1935, she was to spend no less than eight periods away from home waters. Between 1-3.35 she accompanied the Home Fleet on its cruise to the West Indies — the last care-free deployment, which was followed six months later by detachment to the Mediterranean Fleet at Alexandria until 3.36 during the Abyssinian crisis.

On the outbreak of the Spanish Civil War in 7.36, ESK saw considerable service in Spanish waters, being stationed 5-7.36 at Gibraltar, 9-11.36 off the North Spanish ports, 1-3.37 off the South Spanish ports, 4.37 off Republican ports, 10-12.37 again at Gibraltar, 1-3.38 off Southern Spanish ports and finally 1-3.39 based at Gibraltar.

During the Munich crisis of 9.38 ESK and EXPRESS were temporarily under the control of the 9th D.F. at the Nore, undertaking minelaying practice on 3.10.38.

On 24.6.39 ESK started to reduce to Reserve, as manning problems precluded her intended use as a Boys' Training Ship. However, she re-commissioned on 2.8.39 and was present at the Reserve Fleet Review three days later. The imminence of War, after the signing of Soviet-German pact, meant that ESK was converted to her war-time role of a minelayer between 23.8-7.9.39. She joined the 20th Destroyer (M/L) Flotilla the next day and served with this unit until her loss a year later.

After escorting the battleship ROYAL SOVEREIGN from Scapa to Portsmouth between 23-26.9.39, ESK operated from Milford Haven on minelaying duties until 11.39. This was followed by a month's operations from Portsmouth. A brief refit at Portsmouth between 29.12.39-26.1.40 was followed by six months' arduous work as escort for the minelayers PRINCESS VICTORIA and TEVIOTBANK laying the East Coast barrage, as well as individual minelaying operations.

ESK was heavily engaged in the Dunkirk evacuation transporting 3,904 troops back to the U.K. between 29.5-3.6.40. On 1.6.40 she rescued over a thousand French troops from the sinking personnel ship SCOTIA off Dunkirk. After minor repairs for damage sustained in the operation, ESK resumed her minelaying duties.

On the night of 31.8/1.9.40 ESK and her flotilla mates INTREPID, ICARUS, IVANHOE and EXPRESS left Immingham to undertake a minelaying operation off the Dutch coast. When some 40 miles north-west of Texel, EXPRESS was mined at 23.07 hours losing her bow, and ESK closed on the stricken vessel. At about 23.12 hours an explosion occurred forward of ESK. About fifteen minutes later a second large explosion occurred amidships. The vessel appeared to break into two and sank immediately with the loss of 127 of her crew in position 53° 26' 36" N 03° 48' 00" E.

EXPRESS after being mined off Texel on 31.8.40. She was to be in dockyard hands for 14 months, but survived the war as the Canadian GATINEAU. *(Imperial War Museum A534)*

EXPRESS/H.M.C.S. GATINEAU (H61)

EXPRESS nearly came to a premature end, when, on 31.8.40, she was one of a group of destroyers laying a defensive minefield some 40 miles off Texel. At 23.07 hours that day in position 53° 25′ N 03° 48′ E, she exploded a mine abreast of 'B' gun. The whole of her structure forward of stations 38/52 disappeared and 4 officers and 53 ratings were killed and 9 others were subsequently made prisoners of war. At 01.40 hours the next day (1.9.40) she was able to work up to 90 r.p.m. astern and for several hours travelled stern first towards the U.K. However, at 08.40 hours the destroyer KELVIN took EXPRESS in tow until 11.12 hours when the tow parted. At 11.15 hours JUPITER took over the tow and EXPRESS finally arrived at Hull stern first at 17.30 hours on 2.9.40. The rebuilding of her forward structure was not completed at Chatham until 4.10.41.

EXPRESS had originally commissioned on 6.11.34 as a unit of the 5th D.F. She remained in Home Waters for nine months, with her gun mountings being adjusted at Sheerness between 13.12.34-5.1.35. During the Abyssinian crisis, she was attached to the Mediterranean Fleet at Alexandria between 9.35-3.36. She then refitted at Portsmouth between 23.3-4.5.36 before being detached to Gibraltar for the next two months. After a further six months service in Home Waters, she undertook two months' duty on the Non-intervention patrol off the Spanish Mediterranean coast between 1-3.37. On her return to the U.K., she had a brief period of repairs, before undertaking a two month period of minelaying trials.

EXPRESS in pre-war days with A and Y mountings landed and with a full complement of mines. *(Wright & Logan)*

EXPRESS had just completed a refit at Portsmouth between 9.8-2.10.37, when a fire in her No. 1 boiler room resulted in considerable damage to her electrical cabling and further repairs at Gibraltar between 24.10-3.12.37. EXPRESS then operated in Home Waters during 1938 with one spell of patrol duties based at Gibraltar. She operated as a minelayer at Portsmouth between 15.8-4.10.38.

After a refit at Portsmouth between 21.11.38-16.1.39, EXPRESS operated from Gibraltar until 3.39. On 21.3.39 she escorted the ferry COTE D'AZUR convoying the French President on his official visit to the United Kingdom.

On her relief by JANUS in the 5th D.F., EXPRESS was earmarked for conversion to a Boys' Training Ship and aircraft co-operation vessel between 6-8.39. This was not proceeded with as there was insufficient crew available. However, on 2.8.39 she was present at the Reserve Fleet Review. She finally arrived at her war station — Immingham — on 8.9.39 and joined the 20th Destroyer (M/L) Flotilla on its formation and operated with this flotilla until mined a year later.

On the night of 9/10.9.39, EXPRESS laid her first offensive minefield at the suspected exit from the German mine barrage in the North Sea. Other major minelaying operations were the laying of 240 mines in the Ems estuary with ESK, INTREPID and IVANHOE on the night of 17/18.12.39. On 15.5.40 EXPRESS, ESK and IVANHOE laid a barrage of 164 mines off the Hook of Holland (the German minesweepers M61, M89 and M136 were to sink on 26.7.40 on this barrage). However, before this last operation, EXPRESS had been in collision with a trawler on 22.3.40 and repairs at Hartlepool took until 24.4.40.

EXPRESS was then deeply involved in the Dunkirk evacuation, being the second to last vessel to leave that port on the completion of her sixth trip. She evacuated a total of over 3,500 troops.

The mining on 31.8.40 was to be the turning point of EXPRESS's career, as following her post repair trials, she was assigned with ELECTRA the task of escorting the battleship PRINCE OF WALES to the Far East. The squadron left the Clyde on 25.10.41 and sailed via the Cape to join REPULSE with the destroyers JUPITER and ENCOUNTER. EXPRESS was the only one of the six vessels of the squadron to survive by 3.42.

When the PRINCE OF WALES and REPULSE were sunk by Japanese aircraft off the north-east coast of Malaya on 10.12.41, EXPRESS rescued nearly a thousand of the 2,081 survivors of the two vessels.

EXPRESS was then utilised as an escort for "China Force", during 1.42 and operated between Singapore and Java. She was to survive the Java Sea and Trincomalee debacles because on 6.2.42 she suffered a fire in her boiler room and much of her electrical cabling, oil fuel tanks, and bulkheads required re-conditioning. Despite this damage, she operated for several weeks before she repaired and refitted at Simonstown between 25.4-26.6.42.

She then operated as a vessel of the Eastern Fleet's 12th D.F. The principal action at this time was her participation as an escort for the aircraft carrier ILLUSTRIOUS that was providing air cover for the forces that landed at Majunga on 10.9.42 to occupy the remainder of Madagascar.

On her return to the U.K. during 2.43, EXPRESS refitted at Liverpool between 12.3-2.6.43. The next day she commissioned in the Royal Canadian Navy as GATINEAU and twelve days later she was presented to Canada as a gift.

GATINEAU operated as a member of Escort Group C3 on the North Atlantic for the next eleven months. On 6.3.44 she participated in the sinking of U744, whilst escorting convoy HX280.

In May 1944, GATINEAU was transferred to the 11th Escort Group based at Londonderry for invasion duties until 7.44. She then returned to Halifax to refit. This refit, which included a complete re-tubing of her boilers, was completed between 3.8.44-16.2.45.

She then worked up at Tobermory during 3.45 before operating in U.K. waters until V.E. Day. On her return to Canada the next month she repaired at Halifax between 11-19.7.45 before transferring to the West Coast where she paid off into Reserve at Esquimalt on 10.1.46. She was removed from the Canadian Navy List on 1.4.47 and sold to Capital Iron & Steel Metals Ltd., Victoria and is reported to have been scuttled at Royson, British Columbia during 1948 for use as a breakwater.

THE 1932 PROGRAMME

THE LEADER: FAULKNOR

There was little discussion over the design of this vessel, as she was essentially a repeat of EXMOUTH. The Admiralty invited tenders on 15.11.32, with tenders being received by 19.12.32. She was finally ordered from Messrs. Yarrow's on 17.3.33 at the contract price of £271,886* made up of hull £118,300, machinery and boilers £145,351 and auxiliary machinery £8,235.

On 28.6.35 A. P. Cole of the destroyer section reported that FAULKNOR although of the same design as EXMOUTH, was, when inclined, some 36 tons lighter than her sister. It seems that her machinery was 34 tons lighter at 506 tons and her hull 10 tons lighter at 731 tons compared with EXMOUTH's. FAULKNOR in standard condition was 1,458 tons when inclined and not 1,505 as calculated. EXMOUTH's weights had been 1,494 tons as inclined and 1,515 tons as calculated.

*Excluding Admiralty supplied items such as guns.

THE DESTROYERS: THE FEARLESSES

In a memo dated 17.6.32, the Controller gave the following outlines on future destroyer design policy:
(i) All future Flotilla leaders and destroyers to be fitted with ASDIC.
(ii) The destroyers to be fitted with T.S.D.S. or as minelayers as appropriate. However, the conversions were to be completed at short notice.
(iii) No further vessels to be fitted as minelayers, until those minelaying destroyers of the 1931 Programme had had their trials.

Thus the 1931 Flotilla consisted of one Flotilla Leader and 6 destroyers fitted with ASDIC/T.S.D.S, whilst two destroyers were to be fitted with ASDIC, T.S.D.S. and as a minelayer. The 1932, 1933 and 1934 programme Flotillas would be fitted with T.S.D.S. and ASDIC.

However, by 26.10.32, owing to improvements in boiler-design and machinery, it was found practicable to reduce the specified machinery weights for the 1932 Programme destroyers and leader to 515 and 540 tons respectively. The machinery weight of the leader turned out to be 506 tons. Otherwise the vessels were to be repeats of the ECLIPSE class.

Tenders for the vessels were invited on 15.11.32 to be returned by 19.12.32. The successful tenders were:

NAME	BUILDER	TENDER PRICES			TOTAL
		HULL	MACHINERY	AUX. MACHINERY	
FEARLESS	CAMMELL LAIRD	£103,880	£133,312	£8,536	£245,728
FORESIGHT		£103,580	£133,312	£8,536	£245,428
FORTUNE	JOHN BROWN	£105,200	£134,250	£8,114	£247,564
FOXHOUND		£104,870	£134,250	£8,114	£247,234
FORESTER	J. S. WHITE	£105,500	£135,332	£8,066	£248,898
FURY		£105,140	£135,332	£8,066	£248,538
FAME	PARSONS MARINE STEAM TURBINE CO. LTD.*	£104,000	£131,744	£8,472	£244,216
FIREDRAKE		£103,750	£131,744	£8,472	£243,966

* Hulls sub-contracted to Vickers-Armstrongs Ltd. (Tyne)
The contracts for the vessels were placed on 17.3.33.

WEIGHT PROBLEMS:

The weight problems that were to cause great concern in the G, H and I classes made their appearance with the FEARLESSES. When FORTUNE was inclined on 24.4.35, her calculated standard displacement was 1,403 tons and not the 1,350 tons standard displacement quoted for the Class. The estimated G M at deep displacement was 2.2 feet. No other details of the Legend of Particulars of the FEARLESS class survive in Cover 515.

BUILDING DETAILS

NAME	LAID DOWN	LAUNCHED	COMPLETED
FAULKNOR	31.7.33	12. 6.34	24. 5.35
FEARLESS	17.7.33	12. 5.34	19.12.34
FORESIGHT	31.7.33	29. 6.34	15. 5.35
FORTUNE	28.7.33	29. 8.34	27. 4.35
FOXHOUND	15.8.33	12.10.34	21. 6.35
FORESTER	15.5.33	28. 6.34	19. 4.35
FURY	19.5.33	10. 9.34	18. 5.35
FAME	5.7.33	28. 6.34	26. 4.35
FIREDRAKE	5.7.33	28. 6.34	30. 5.35

FAULKNOR

After commissioning on 27.5.35, FAULKNOR worked up, but defects kept her at Portsmouth between 29.7-21.9.35, whilst FIREDRAKE deputised as leader. She was then leader of the 6th D.F., renumbered the 8th in 4.39, until 10.44. FAULKNOR remained in the Mediterranean until 7.36, before refitting at Portsmouth between 20.7-3.10.36. After a period in Home Waters, she spent between 1-3.37 in Spanish Mediterranean waters and then a further three months in the Bay of Biscay.

At 06.20 hours on 4.8.37, FAULKNER collided with the freighter CLAN MACFADYEN off Ushant, receiving considerable damage, which was repaired at Portsmouth between 5.8-28.12.37. She then operated from Gibraltar with the French Navy between 1-3.38, before spending the remainder of her pre-war service in Home Waters.

On the outbreak of war, FAULKNOR was leader of the 8th D.F. of the Home Fleet. She was soon to see action whilst screening the aircraft carrier ARK ROYAL, as FAULKNOR with FOXHOUND and FIREDRAKE sank U39 by depth charges and gunfire in position 58°29'N 11°50' W, after the submarine had attempted to torpedo ARK ROYAL. The entire crew of 43 were rescued. On 12.2.40 FAULKNOR rescued 10 survivors from the Swedish vessel ORANIA six miles north-east of the Shetlands. ORANIA had been sunk the previous day by U50.

On 15.4.40, FAULKNOR engaged a German merchant ship and gun emplacements in Narvik Harbour. The next day FAULKNOR and ZULU bombarded a wrecked destroyer in Herjang Fjord. A boarding party from FAULKNOR on the German destroyer was fired upon by snipers and suffered a fatal casualty. FAULKNOR then survived two bombing attacks on the 16.4.40 and 20.4.40.

On 5.5.40, FAULKNOR struck a wreck in Rombak Fjord in position 68°27'N 17°38'E and her ASDIC dome and directing gear were put out of action. She returned to the U.K. and refitted and repaired at Doig's of Grimsby between 11.5-13.6.40. Four days later she left Scapa for Gibraltar to join Force 'H' on 21.6.40.

FAULKNOR operated with Force 'H' until 1.41, participating in escort duties, returning briefly to Liverpool 11-19.8.40 for repairs before participating in Operation 'MENACE' — the attack on Dakar. Between 15-19.11.40 she was one of the escorts for the aircraft carrier ARGUS undertaking a flying-off operation to Malta. Three days later, FAULKNOR and FORESTER intercepted the French motor vessel CHARLES PLUMIER some 110 miles east of Gibraltar and brought her into Gibraltar.

FAULKNOR was then detached for escort duties at Freetown between 1-3.41, before returning to her duties with Force 'H' at Gibraltar between 3-8.41. On 3.2.41 she rescued four men from the British tanker BRITISH PREMIER (sunk by U65 off Freetown). The men had spent 41 days in an open boat. Soon after returning to Gibraltar, FAULKNOR with FORESTER, FORESIGHT and FEARLESS sank U138 in position 36°04'N 07°29'W off the Straits of Gibraltar on 18.6.41. 28 survivors were picked up.

During 8.41 FAULKNOR suffered turbine damage which on inspection was found to have been caused by sabotage. Repairs were completed at Southampton between 20.8-13.11.41. FAULKNOR then arrived at Scapa on 22.11.41 to work up, but was later damaged when proceeding alongside an oiler and temporary repairs were completed by the depot ship TYNE. She then operated with the Home Fleet until 6.43.

She operated in Arctic waters for the whole of this period, except between 1.4-4.4.42, when she was one of a group of vessels that attempted to cover the passage of ten Norwegian ships from Gothenburg to the U.K. Only two vessels — the B.P. NEWTON and LIND reached the U.K., whilst six others were sunk and the remaining two vessels returned to Sweden. On 2.4.42, FAULKNOR sank the damaged RIGMOR by torpedo and gunfire before the force withdrew. She refitted at Amos and Smith at Hull between 13.7-28.8.42.

After working-up during 9.42, FAULKNOR joined the escort of Convoy PQ18, and participated in sinking her third U-boat (U589) by depth charges south-west of Spitzbergen. She again refitted at the yard of Amos and Smith at Hull between 17.3-16.4.43. She then operated in Icelandic waters until ordered to the Mediterranean to participate in the Sicily landings. She left Scapa on 17.6.43 and arrived at Alexandria on 5.7.43. She then bombarded Locri, north-east of Cape Spartivento, on 28-29.7.43.

FAULKNOR on 14.1.42 as leader of the Home Fleet's 8th D.F. Note the 3" H.A. in lieu of the after bank of torpedo tubes and augmented light A.A. armament.

FAULKNOR then operated in the Eastern Mediterranean with her flotilla until 3.44, before transferring to Gibraltar for the next two months. This period of her service was to see FAULKNOR in action many times. On 17-18.9.43, FAULKNOR with ECLIPSE and the Greek destroyer QUEEN OLGA sank a merchant ship and a tanker off Stampalia. Two days later, she ran troops and supplies to Leros.

On 7.10.43, the cruisers SIRIUS and PENELOPE with FAULKNOR and FURY as escorts, intercepted an enemy convoy in the Scarpanto Strait, sinking six landing craft, an ammunition ship and and an armed trawler. A fortnight later FAULKNOR and PETARD were attacked by enemy aircraft using flares in Parthani Bay. FAULKNOR escaped damage, but PETARD was shaken by a near miss. Between 2-6.11.43, FAULKNOR with ECHO, PENN and PATHFINDER landed a battalion of 750 men and 80 tons of stores at Leros. Five days later, FAULKNOR with BEAUFORT and PINDOS bombarded Kos harbour and between 13-14.11.43, they bombarded enemy positions on Leros following the successful German landing.

FAULKNOR's next operation was to escort the transports ROYAL ULSTERMAN and PRINCESS BEATRIX, which were ferrying 9 Commando, raiding the mouth of the Garigliano river in Italy. FAULKNOR, LAFOREY and the Dutch FLORES bombarded Gaeta and Itri as a diversion. On 18.1.44, FAULKNOR supported the 5th Army's advance across the Garigliano River.

FAULKNOR returned to the U.K. for escort duties during the 'D' Day landings. She then repaired on the Humber between 8.7-21.7.44 and on 26.7.44 she destroyed one of 12 Ju 88's that were attacking offshore patrols in which FAULKNOR was participating. She undertook escort duties in the Channel between 10-12.44 on attachment to the 14th Escort Group, based at Milford Haven. In 12.44 on her return to the 8th Flotilla she operated from Plymouth until the end of the European war. She participated in the blockade of the German occupied Biscay ports.

She was laid-up at Dartmouth on 12.6.45, entering Category 'B' Reserve on 25.7.45. She was then relegated to Category 'C' Reserve on 27.12.45. On 22.1.46 she was made available to BISCO and handed over to T.W. Ward Ltd. at Milford Haven on 4.4.46 for demolition.

FAME (H78)

After commissioning, FAME joined the 6th D.F. of the Home Fleet (the 8th after 4.39) until the outbreak of war. However, she had to return to Devonport for modification of her ammunition hoists between 23.7-28.8.35. FAME then joined the Mediterranean Fleet until 7.36 for the duration of the Abyssinian crisis. On her return to the U.K. she refitted at Devonport 20.7-10.11.36, before service in Spanish waters until 1.37. She served again off the Biscay ports between 8-9.37 and then undertook patrols based at Gibraltar and Oran with the French Fleet for the next month. She visited Aarhus, Denmark between 12-18.7.37 and then spent the remainder of her pre-war service in Home Waters.

On the outbreak of war, she remained with the 8th D.F. on patrol and escort duties until 7.40. She served in Norwegian waters, receiving slight damage from gunfire at Narvik on 13.4.40.

On 6.7.40, whilst searching for the submarine SHARK, she was attacked by German aircraft off the east coast of Scotland and received severe splinter damage. She commenced repairs by Robbs at Leith, until 26.9.40, completing at Rosyth by 10.10.40.

On 17.10.40, whilst escorting the new battleship KING GEORGE V to Rosyth, she ran aground on the Northumberland coast and was severely damaged in the swell and by an electrical fire. The ship had to be lightened and she was not refloated until 1.12.40.

FAME on 5.9.42 on completion of her rebuilding and conversion to an A/S destroyer following her grounding on the Northumbrian Coast on 17.10.40.

Temporary repairs were undertaken at Sunderland until 2.41 and FAME finally arrived in tow at Chatham on 5.2.41 for rebuilding. These works were not completed until 9.42 and included her conversion to an escort destroyer.

On commissioning, she joined Escort Group B6 of the Western Approaches Command and was soon in action, as on 16.10.42, whilst escorting convoy SC104 in the North Atlantic, she rammed and sank U353, picking up 39 survivors. During 12.42, FAME with the remainder of Escort Group B6, protected convoy HX217 for four days against the attacks of no fewer than 33 U-boats, for the loss of only two vessels of the convoy.

FAME continued on escort duties until 4.44, sinking her second submarine (U69) with no survivors north-east of Newfoundland on 17.2.43, whilst protecting ONS165. During this month, FAME transferred to the 14th Escort Group, with which she remained for the remainder of the war. She participated in the Normandy landings and sank her third U-boat — U767, with INCONSTANT and HAVELOCK south-west of Guernsey on 18.6.44. On the conclusion of operation 'Overlord' she was employed on escort duties on the west coast of Scotland.

Refitted at Leith between 5.45-8.45, she joined the Rosyth Escort Force until 10.45, when she was transferred to the Londonderry Training Flotilla. However, a month later, she became senior officer's ship of the 3rd Flotilla at Londonderry. She served with this flotilla until 10.46 at Londonderry and then at Portland to 4.47. She entered Reserve during 5.47 at Portsmouth. On 20.1.48 she was classified as Category 'C'. She left Reserve during 6.48 and was immediately re-fitted.

On 4.2.49, she was transferred at Devonport to the Dominican Republic as GENERALISIMO. The total cost of FAME and HOTSPUR together was £190,000 plus £40,000 each to refit and £39,000 for armament stores and £28,000 for naval stores. She served in the Dominican Navy until 1968 when she was discarded, having been re-named SANCHEZ in 1962 following a revolution.

FEARLESS (H67)

On completion FEARLESS joined her sisters in the 6th D.F., and served with the Home Fleet for all her pre-war service but was detached to the Mediterranean Fleet between 3-7.36 and served off the Biscay Ports 11.36-1.37 and 8-10.37 and in the Gibraltar area 2-3.37, and 1-3.38 and 1-3.39 and escorted the s.s. HABANA with refugee Basque children aboard into St. Jean de Luz.

On 3.9.39, FEARLESS was stationed with the by now 8th D.F. of the Home Fleet at Scapa, where she remained based for the next two months. On 7.9.39 she arrived at Haugesund to pick up the Polish Mission in order that the Mission would not be interned. On 20.9.39, FEARLESS was one of the four destroyers that sank U27 off the Hebrides. She then resumed her duties with the Home Fleet until 14.5.40. A month earlier on 14.4.40 FEARLESS and BRAZEN had sunk U49 off Vaagsfjord with depth charges. FEARLESS was again under repair between 15.5-10.6.40 at Middlesbrough. Repairs were effected to a rudder gland and to boiler defects. A week later FEARLESS with her sisters FAULKNOR and FOXHOUND and the destroyer ESCAPADE escorted the battle-cruiser HOOD and aircraft carrier ARK ROYAL between Scapa and Gibraltar, where they arrived on 23.6.40. FEARLESS escorted Force 'H' during operation 'CATAPULT' — the attack by British Naval Forces on the French Navy at Mers-el-Kebir, Oran on 3.7.40 which resulted in the loss

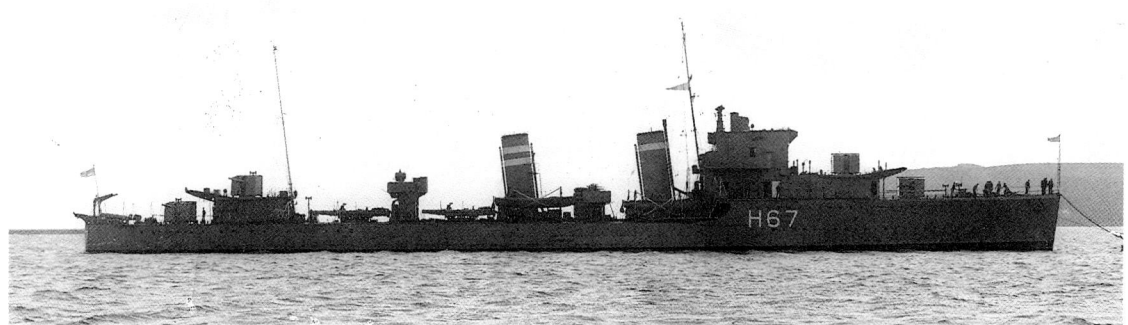
FEARLESS on completion in 1935. She was little altered at the time of her loss on 23.7.41.

of the French battleship BRETAGNE and the damaging of the battle-cruiser DUNKERQUE. Six days later she participated in the inconclusive action off Calabria. On 2.8.40 she acted as escort when the ARK ROYAL's aircraft attacked Cagliari on Sardinia. This ended an action-packed period for FEARLESS as she then returned to the U.K. on 4.8.40, but on her passage there, she collided with the trawler FLYING WING. Collision repairs by Barclay Curle at Scotstoun were effected between 10.8-11.10.40. However, she had hardly re-entered service when she was in collision with the s.s. LANARK at anchor at Greenock on 30.10.40 and her stern was fractured. Repairs this time were undertaken at Troon and were not completed until 8.1.41, the vessel having been taken in hand on 10.11.40.

On completion of repairs and a brief work-up, FEARLESS returned to the Gibraltar based 8th D.F. on 18.1.41 with Force 'H' and operated with this flotilla until her loss six months later. During this period she participated in the attack on Genoa on 6.2.41, and attempted the seizure of the French merchant ship BANGKOK, alleged to be carrying 3,000 tons of rubber, but was thwarted by coastal gun-fire from a battery at Nemours, which forced FEARLESS to withdraw on 30/31.3.41. FEARLESS acted as escort for Operation 'WINCH', a flying-off operation to Malta on 3.4.41.

Between 5-12.4.41, FEARLESS and the other members of the 8th D.F. escorted the ships of Operation 'TIGER' through to Malta and used their minesweeping equipment to clear the channel into Malta so that the blockaded 5th Flotilla (Capt. Lord Mountbatten) could emerge. The destroyers refuelled and then escorted Force 'H' to Gibraltar. A month later, on a sweep of the Atlantic with Force 'H', FEARLESS, FAULKNOR, FORESIGHT and FOXHOUND sank U138 on 16.6.41.

FEARLESS' career came to an end during Operation 'SUBSTANCE' a supply operation to Malta escorted by Force 'H' and elements of the Home Fleet. On the third day out (23.7.41), FEARLESS was hit by aircraft bombs off Bone, Algeria, and her stern set on fire. FORESTER took off survivors and FORESIGHT sank FEARLESS by torpedo. 1 officer and 15 ratings were listed missing, whilst nine ratings died of wounds and 11 others were wounded. The vessel lies in 37° 40'N 08°20'E.

FIREDRAKE (H79)

On completion on 30.5.35, FIREDRAKE joined her sisters in the 6th D.F. of the Home Fleet, which was re-numbered the 8th during 4.39. After a few months in Home Waters, FIREDRAKE operated from Gibraltar between 9-12.35. She acted as leader between 29.8-24.9.35. A refit at Gibraltar between 14.12.35-11.2.36 was followed by a brief spell in Home Waters to give leave before she was attached to the Mediterranean Fleet 3-7.36.

FIREDRAKE survived severe shock damage in the Mediterranean on 23.7.41 only to be lost by torpedo attack in the North Atlantic on 17.12.42.

She again refitted at Sheerness between 23.7-30.9.36 and began non-intervention patrol duties in Spanish waters until 6.37. On 19.4.37, FIREDRAKE supported by the battle-cruiser HOOD, escorted the steamer MACGREGOR into Bilbao from the attentions of the Spanish Nationalist cruiser ALMIRANTE CERVERA and the armed trawler GALERNA. She returned to Gibraltar for two months of patrol duties from 9.37. On 8.10.37 she assisted the steamer CERVANTES that had allegedly been bombed. FIREDRAKE then refitted at Sheerness between 3.11.37 and 30.12.37.

During the ensuing period before the outbreak of war FIREDRAKE saw further service in Spanish waters between 1-3.38 when based at Gibraltar and a year later when she served on patrol duties in the Bay of Biscay.

On 3.9.39 she was lying at Scapa and eleven days later participated with FAULKNOR and FOXHOUND in the sinking of U39 (see FAULKNOR). On 25.9.39 FIREDRAKE participated in the rescue operation for the submarine SPEARFISH, badly damaged whilst on patrol in the Heligoland Bight.

FIREDRAKE spent the next six months on normal destroyer duties with the Home Fleet. On 4.10.39 she rescued six survivors from the steamer GLEN FARG some 25 miles east of the Orkneys. She was damaged whilst going alongside ICARUS at Invergordon on 28.3.40 and spent between 2-26.4.40 under repair at Cardiff.

FIREDRAKE was then engaged on operations in Norwegian waters, escorting convoys between Narvik and Harstad, bombarding enemy positions in Ofot Fjord and participating in the withdrawal of troops from Bodo and Harstad. She was slightly damaged by splinters on 23.5.40 and again on 12.6.40 when her port steering motor and 'A' gun were put out of action. Repairs on the Clyde were completed on 20.6.40.

Still based at Scapa, FIREDRAKE joined the 4th D.F. briefly, before leaving Scapa on 22.8.40 as one of the escorts for the aircraft carrier ILLUSTRIOUS and the cruisers SHEFFIELD and YORK on passage to Gibraltar. Meanwhile on 2.7.40 she participated in the search for survivors of the liner ARANDORA STAR sunk by U47 (Prien) some 125 miles off Malin Head.

On 29.8.40 she arrived at Gibraltar, where she was to be based for the next six months, a period which proved to be exceptionally busy for the 8th D.F. of Force 'H'. On 18.10.40 FIREDRAKE with the destroyer WRESTLER and two London flying boats, sank the Italian submarine DURBO in position 35°57'N 04°00'W east of Gibraltar. Documents retrieved from the submarine gave the disposition of the Italian submarines and two days later this information resulted in the sinking of the submarine LA FOLE (see GALLANT).

On 27.11.40 FIREDRAKE participated in the action off Cape Spartivento and on 1.1.41, with the destroyers DUNCAN, FOXHOUND, HERO and JAGUAR intercepted four Vichy French ships and escorted them to Gibraltar. Later the same month, she participated in Operation "EXCESS" — a supply convoy to Malta and with the remainder of Force 'H' bombarded Genoa.

However, on 1.3.41 FIREDRAKE went aground in fog near Gibraltar in position 36°28'N 04°45'W, losing her ASDIC dome and damaging her propellers. Repairs were undertaken at Gibraltar until 21.4.41 and completed at Chatham by 19.6.41. She then returned to her duties with the 8th D.F. of Force 'H', after escorting the 9th M.L. Flotilla to Gibraltar. She was not to remain on station long because at 17.15 hours on 23.7.41 she was narrowly missed by a 100kg bomb, which exploded close to the starboard side near No. 1 boiler room. Her side plating was blown inwards from the upper deck to bilge keel for most of the length of her boiler room. FIREDRAKE was escorted by the destroyers AVON VALE, ERIDGE and SIKH into Gibraltar, where she arrived on 27.7.41. Throughout 1940/41 FIREDRAKE operated with a split in the boiler room casing.

Temporary repairs were undertaken at Gibraltar and on 13.9.41 in the company of the cruiser MANCHESTER, FIREDRAKE left for permanent repairs at the Boston Navy Yard. These repairs were effected between 23.9.41 and 12.1.42 and included the removal of one 4.7" gun and the fitting of additional depth charges to suit her for her new duties as an A/S escort with the Western Approaches Command.

After escorting convoy NA2 to the U.K., FIREDRAKE joined the escort group B7 with which she served until her loss and was immediately employed escorting convoy WS 16. Repairs to her ASDIC dome were undertaken on the Clyde during 4.42. More Atlantic convoy duties followed and further repairs were undertaken at Belfast during 8.42. On 26.9.42 she picked up survivors from the OLAF FOSTENES, which had been sunk eight days earlier.

At 19.11 hours on 16.12.42 whilst escorting convoy ON153 and when some 550 miles west of Cape Clear, FIREDRAKE was torpedoed by U211. The torpedo struck her on the starboard side of the forward boiler room. She broke in two and the bow section floated away and sank. The stern section aft of the boiler room remained afloat until 00.45 hours on 17.12.42, when it sank. 26 survivors from her complement of some 140 were rescued by the corvette SUNFLOWER. The vessel lies in 50°50'N 25°15'W.

FORESIGHT (H68)

FORESIGHT was a member of 6th D.F. of the Home Fleet from commissioning until 4.39, when the Flotilla was renumbered the 8th. She was then to operate with this Flotilla for the first eight months of the war. In 5.40 she was detailed to the east coast, based on the Humber for the evacuation of Dunkirk and later from the other Channel ports.

A month later, she was one of the original vessels sent to Gibraltar to form Force 'H', and she was to remain with the 8th D.F. until 10.41. She participated in the abortive landing at Dakar, sinking the French submarine BEVEZIERS on 25.9.40, after the latter had torpedoed and damaged the battleship RESOLUTION off that port. FORESIGHT then returned to the U.K. to refit at Liverpool where she was damaged by a near miss during an air-raid on the night of 21/22.12.40. After further repairs, she rejoined Force 'H' and participated in many actions over the next ten months — the bombardment of Genoa 9.2.41, two flying-off operations to Malta during 4.41, as an escort for the 'TIGER' convoy to Malta during 5.41 and a flying-off operation to Malta the next month.

On 18.6.41 FORESIGHT, with her sisters FAULKNOR, FEARLESS, FORESTER and FOXHOUND sank her second submarine (U138) in the Straits of Gibraltar. A month later, on 23.7.41, FORESIGHT had the task of sinking her sister FEARLESS, during Operation 'SUBSTANCE'. Further escort operations to Malta followed during 7.10.41.

During 10.41 the 8th D.F. returned to the U.K. and were escorts for the Home Fleet after undertaking escort duties in the Bay of Biscay. FORESIGHT was then to undertake escort duties for Russian convoys. On 2.5.42 FORESIGHT and FORESTER had been towing and escorting the damaged cruiser EDINBURGH from Kola to Iceland when they were attacked by German destroyers. During the action FORESIGHT was hit four times. The first shell burst on impact with the ship's side plating abreast the Engineer Officer's cabin. The second shell burst on impact with the ship's side plating abreast the torpedo-tubes between stations 56/59. Oil was lost from three out of the four fuel tanks. The third hit caused a 4ft hole in the deck plating and splinters damaged the boiler room and the ship was immobilised and all electrical power ceased. The fourth hit perforated the radar hut and burst on contact in the drying room. FORESIGHT lost 8 killed and 11 wounded in the action.

EDINBURGH had been torpedoed again during the action and FORESIGHT and FORESTER took off her survivors and scuttled her.

After a brief refit during 6.42 and further escort duties to Russia, FORESIGHT was one of the Home Fleet vessels attached to the Mediterranean Fleet to cover the 'PEDESTAL' convoy to Malta.

FORESIGHT pictured on 2.7.42 following her refit after arduous Arctic convoy duties. She is still armed as a fleet destroyer.

The derelict FORESIGHT some six weeks later, on 13.8.42, after being hit by an aircraft torpedo in the Sicilian Narrows.
(Imperial War Museum HU47957)

On 12.8.42, whilst still in the Sicilian Narrows, FORESIGHT was hit by an aircraft torpedo and disabled. She was towed towards Gibraltar by the destroyer TARTAR, but had to be sunk the next day off Galita Island due to the presence of enemy forces. One officer and three ratings were presumed killed.

FORESTER (H74)
On 3.9.39 FORESTER was at Scapa with other vessels of the 8th D.F. She had been completed in 4.35 and immediately joined the 6th Flotilla remaining with this flotilla until it was renumbered. However, she spent several periods on detachment — at Gibraltar between 9.35-2.36 and again at Gibraltar on patrol duties 9-10.37.

FORESTER spent the first nine months of the war on fleet escort duties based at Scapa. The highlight of this period was FORESTER's sinking of U27 west of the Hebrides by depth charges and gunfire on 20.9.39. In this she was aided by her sister FORTUNE. Some five months later on 11.2.40, she and the tug BUCCANEER towed into port the tanker IMPERIAL TRANSPORT which had been torpedoed some 150 miles north-west of the Butt of Lewis. Later the same month she escorted the liners ORION and DUCHESS OF BEDFORD transporting elements of the R.C.A.F. to the United Kingdom. On 13.4.40 she escorted the battleship WARSPITE during the second battle of Narvik.

On 26.6.40 FORESTER left Scapa for Gibraltar to join Force 'H' with which she operated until 26.10.41. This period was one of hectic activity, as she escorted five Mediteranean or Malta convoys — Operation 'COLLAR' during 11.40, Operation 'TIGER' during 5.41, 'SUBSTANCE' and 'STYLE' in 7.41 and 'HALBERD' in 9.41. It was during the 'SUBSTANCE' Operation that she rescued the survivors of her sister FEARLESS, damaged by aerial torpedo on 23.7.41 and which had subsequently to be sunk.

FORESTER also participated in the disastrous attack on Dakar during 9.40 and escorted the aircraft carriers FURIOUS and ARK ROYAL during flying-off Operations to Malta in Operation 'SPLICE' in 5.41 and Operation 'TRACER' on 15.6.41, Operations 'STATUS' and 'STATUS II' when 49 Hurricanes were flown to Malta in 9.41 and finally Operation 'CALLBOY'', when Albacore torpedo bombers were flown to Malta a month later.

On 18.6.41 when returning from Operation 'TRACER', FORESTER with FEARLESS, FAULKNOR, FORESIGHT and FOXHOUND sank U138 in the Straits of Gibraltar.

On her return to the U.K. on 26.10.41, FORESTER was temporarily attached to the 11th Escort Group on the Clyde, before returning to the Home Fleet at Scapa the next month. She then refitted as an escort destroyer.

Her refit and work-up completed, she rejoined the Home Fleet and was immediately engaged in severe actions in the Arctic in April and May 1942.

On 2.5.42 FORESTER, FORESIGHT, two mine sweepers and a Russian destroyer were screening the cruiser EDINBURGH, which had been torpedoed by U456 whilst protecting convoy QP11 on 29.4.42 and was returning to Kola Inlet (*see* FORESIGHT for details of the action). During the action FORESTER lost her commanding officer and 12 ratings killed with another 9 ratings wounded. She also received hits in her boiler room and damage to B and X guns.

FORESTER prior to her conversion to an A/S destroyer in 4.43. She retains her searchlight and 3" gun.

After undertaking temporary repairs at Murmansk, FORESTER left there on 13.5.42 in company with SOMALI, FORESIGHT and MATCHLESS as escort for the cruiser TRINIDAD previously damaged whilst escorting convoy PQ13. The next day the group was attacked by aircraft and TRINIDAD was set on fire and had to be abandoned. FORESTER evacuated TRINIDAD's wounded, passengers and crew before TRINIDAD was sunk. FORESTER finally arrived at Scapa on 18.5.42 and immediately proceeded to the Tyne for a refit and repairs that were not completed until 10.42.

On rejoining the Home Fleet's 8th D.F. FORESTER escorted convoys QP15, JW51B, RA52, JW53 and RA53 to and from Russia until 4.43. She then refitted, this time at Leith. This work was completed two months later. During this refit, her light anti-aircraft armament was altered by the fitting of 6 Oerlikons and the suppression of her 3" gun. She then spent the next year on convoy escort duties in the Western Approaches Command, as part of Escort Group C1.

On 10.3.44 FORESTER with ST. LAURENT and the Canadian frigates OWEN SOUND and SWANSEA sank U845 in Mid Atlantic. FORESTER then operated as an escort in the Channel supporting the Normandy Landings between 29.5 and 9.44. On 20.8.44 with the destroyers WENSLEYDALE and VIDETTE, she sank U413 off Beachy Head. FORESTER had earlier been in action with the frigate STAYNER against German R Boats off Cap d'Antifer on 23.7.44.

FORESTER then returned to North Atlantic convoy duty, with the 14th Escort Group based at Londonderry until 12.44. On the first of that month she started repairs at Liverpool, which were not completed until 5.45. After working up, FORESTER joined the Rosyth Escort Force where she remained until 8.45. She reduced to Category 'B' Reserve on 2.11.45 at Dartmouth. She was handed over to BISCO on 22.1.46 and arrived at Metal Industries' Rosyth yard for demolition on 28.2.46.

FORTUNE/H.M.C.S. SASKATCHEWAN (H70)

FORTUNE commissioned on 8.5.35 and immediately joined the Home Fleet's 6th D.F. until 4.39. She served with the Mediterranean Fleet in 1937, before undertaking patrol duties in the Bay of Biscay during the Spanish Civil War. In 4.39 with the entry into service of the "Tribals" and with the pending arrival of the J's and K's, the 6th D.F. was re-designated the 8th D.F., although it continued on fleet duties until the outbreak of the war.

The 8th D.F. was immediately in action and sank two of the first U-boats to be lost. FORTUNE participated in the sinking of U27 on 20.9.39 by depth charges and gunfire off the Hebrides and rescued her crew. Six months later, on 20.3.40, whilst acting as escort for the battle-cruisers of the Home Fleet covering a sweep of cruisers north of the Shetlands, FORTUNE sank U44 by depth charges without survivors. This period of escort duties was ended by the opening of the Norwegian campaign.

The principal operation undertaken was on 8.5.40, when FORTUNE and FEARLESS escorted the cruisers BERWICK and GLASGOW, with the 2nd battalion of the Royal Marines for the allied occupation of Iceland, which was completed on 10.5.40.

On 6.7.40 FORTUNE and three other destroyers were despatched at high speed to search and escort home the submarine SHARK, previously damaged by aircraft bombs and unable to dive. The search under heavy air attack was, however, in vain, as the SHARK had surrendered after a gallant fight, later being lost under tow of German minesweepers in the North Sea.

During 8.40 FORTUNE was briefly allocated to the 4th D.F. She stood by the A.M.C. TRANSYLVANIA torpedoed by U56 on 10.8.40 off the north coast of Ireland, and rescued her survivors. She was transferred to the newly created Force 'H' at Gibraltar on 29.8.40 and she immediately participated in the unsuccessful attack on Dakar on 24.9.40. During the action FORTUNE sank the French submarine AJAX, rescuing her crew of 76, after the submarine had attempted to attack the fleet off Dakar.

FORTUNE spent the next five months on escort duties, between Freetown and Gibraltar, and this was followed by three months service with Force 'H' at Gibraltar. She participated in the flying-off operation to Malta—Operation 'WINCH' with the aircraft carrier ARK ROYAL during 4.41 and Operation 'DUNLOP' with the old aircraft carrier ARGUS during the following month.

On 10.5.41 when FORTUNE with FAULKNOR, FEARLESS and FURY were returning to Gibraltar and were about 20 miles off Bougaroni, Algeria, they were attacked by aircraft. At 12.39 hours, FORTUNE was near missed on the starboard side by station 141 by a 250kg bomb which exploded about 20ft below the surface. The explosion stopped all turbo-driven machinery, ruptured the hull for 12ft between stations 142/143 and flooded compartments between stations 125-155 and the vessel listed to starboard.

The wreckage was cleared away and the after magazine and shell room hatches were then closed and speed reduced to 'slow'. The magazine and shell room hatches had to be re-opened to reduce pressure for fear that the deck would burst. Speed of 12 knots was maintained until her return to harbour. Some 16,000lb of equipment was removed from the upper deck within 20 minutes of the explosion, plus 1,500lb in weight of davits, throwers and towing ropes were also jettisoned. Her shafts were bent and the X and Y guns put out of action due to the weakening of the structure in the vicinity and flooding of magazines.

FORTUNE made passage to the U.K. for repairs at Chatham which were not completed until 11.41. She immediately returned to Gibraltar, but mechanical problems and not being fully worked up, restricted her to local patrol duties until 2.42. On 9.2.42 FORTUNE was part of the escort for a convoy to Malta that left Gibraltar that day. She traversed the Mediterranean, acting as escort for the armed transport BRECONSHIRE between Malta and Alexandria, where she arrived on 17.2.42.

FORTUNE then made passage to join the 2nd D.F. of the Eastern Fleet and arrived at Trincomalee on 7.3.42, spending the next three months on escort duties with the Eastern Fleet. On 6.4.42 she rescued 88 survivors from the motor vessel GLENSHIEL torpedoed some 300 miles east from Addu Atoll.

FORTUNE when serving with the 2nd D.F. in the Indian Ocean during 1942.

In 6.42, FORTUNE with HOTSPUR and GRIFFIN of the 2nd D.F. was attached to the Mediterranean Fleet to participate in the unsuccessful Malta convoy operation 'VIGOROUS'.

FORTUNE then returned to the Eastern Fleet and the 12th D.F. for the remainder of 1942. After making passage round the Cape, FORTUNE started a refit at London in 2.43. On completion of this refit, on 31.5.43, she recommissioned as H.M.C.S. SASKATCHEWAN. She was transferred to Canada as a gift on 15.6.43.

She was immediately assigned as senior ship to Escort Group C3 based at Londonderry, until 5.44.

In 5.44, she joined the 12th Escort Group for Operation 'NEPTUNE' and undertook escort and patrol duties in the Channel after 'D' Day. However, on the night of 5/6.7.44 SASKATCHEWAN with QU'APPELLE and SKEENA participated in Operation 'DREDGER' during which one armed trawler was sunk and two others severely damaged off Brest. The trawlers had been escorting two U-boats, which escaped in the confusion. SASKATCHEWAN suffered 1 killed and 4 wounded.

The 12th Escort Group returned to Londonderry at the end of 7.44 and then SASKATCHEWAN sailed independently for Canada. She arrived at Halifax on 6.8.44 and refitted at Shelbourne until 11.44. Further work had to be undertaken at St John's, Newfoundland and SASKATCHEWAN did not finally return to the U.K. until 1.45. She then participated on escort duties with the 14th Escort Group and later the 11th Escort Group until the end of the war. She was at Plymouth on V.E. Day and returned to Canada from Greenock on 30.5.45, conveying Canadian personnel. She undertook five more trooping voyages taking personnel from St. John's to Quebec City before being declared surplus to requirements on 23.9.45, but did not finally pay off until 28.1.46 at Sydney. Later the same year was sold to the International Iron & Metal Co. of Hamilton for scrap.

FOXHOUND/H.M.C.S. QU'APPELLE (H69)

FOXHOUND spent the whole of her pre-war career, after commissioning on 29.6.35, with the same Flotilla — the 6th of the Home Fleet — renumbered the 8th D.F. in 4.39. However, she operated off Vigo and Corunna between 11.36-1.37 and, after a period for repairs at Sheerness, between 16.1-13.2.37 operated off Gibraltar during 2-3.37 and the Biscay ports between 5-6.37 and again between 8-10.37. FOXHOUND refitted at Chatham between 27.10-30.12.37 and then undertook patrol duties between Gibraltar and Oran during 1-3.38. She then operated in Home Waters for the rest of 1938-39.

On 22.9.38, whilst on passage from Rosyth to Invergordon, FOXHOUND was in collision with the submerged submarine SEAHORSE, FOXHOUND received slight damage to her starboard propeller but proceeded to Invergordon. The propeller was repaired 10.11-12.12.38 at Sheerness.

For the first eight months of the war FOXHOUND operated with the 8th D.F. as a unit of the Home Fleet. The highlight of this period of service was her sinking of U39 (*see* FAULKNOR).

The entire crew of U39 was taken prisoner. Some five months later, on 11.2.40, she rescued the survivors of the Swedish vessel ORANIA torpedoed by U50 some 60 miles north-east of the Shetlands.

FOXHOUND was heavily engaged in the Norwegian Campaign, sinking the German destroyer DIETHER VON ROEDER inside Narvik harbour during the second battle of Narvik on 13.4.40. In 5.40 she escorted the battleship WARSPITE to Gibraltar, returning in time to escort the vessels carrying the occupation force to Iceland. She remained in Icelandic waters for the next month, before making passage to join Force 'H' at Gibraltar and on 3.7.40 she participated in the action at Oran against the French Fleet.

FOXHOUND lying disabled with condenser trouble off Takoradi on 29.11.40. Taken from the cruiser DIDO.
(National Maritime Museum N31648)

After refitting at Sheerness between 8-10.40 FOXHOUND returned to Force 'H' and the 8th D.F. for the next ten months. During this period she participated in three ferry operations to Malta, the bombardment of Genoa on 8/9.1.41 and escort duties between Gibraltar and West Africa. On 18.6.41 she was one of the vessels of the 8th D.F. that sank U138 west of Gibraltar.

FOXHOUND returned to the U.K. during 8.41 to refit, which was completed three months later. After undertaking escort duties she briefly joined the 2nd D.F. in the Mediterranean between 1-3.42. In 4.42 she was one of the reinforcements sent to the Eastern Fleet and served with the 2nd Flotilla on the South African station until 4.43. She then transferred to Freetown and the West African station and served with the 4th D.F. between 5-7.43.

She returned to the U.K. during 8.43 for a long overdue refit and conversion into a long range escort vessel and was then turned over to the Royal Canadian Navy. Conversion completed, FOXHOUND commissioned as QU'APPELLE on 8.2.44 and after working up at Tobermory, she joined the 6th Escort Group of the Western Approaches Command, based at Londonderry. During 4.44 QU'APPELLE joined the 12th Escort Group, as senior ship, patrolling the western approaches to the Channel after 'D' Day.

FOXHOUND as QU'APPELLE after her conversion to an A/S destroyer and commission into the Royal Canadian Navy on 8.2.44. Note the position of Hedgehog. (National Maritime Museum N32276)

On 6.7.44 she was in action off Brest, when three German Flak ships were destroyed and again on 11.8.44 when two trawlers were sunk. During this action QU'APPELLE was struck by SKEENA and was under repair until 5.9.44. During 10.44 she joined the 11th Escort Group and operated off Iceland until 12.44. She was present when SKEENA was lost at Reykjavik on 25.10.44.

She refitted at Halifax between 5.12.44-30.6.45. The war being now over, she ferried personnel home from the U.K. and made four crossings from Greenock before 25.9.45. She was then transferred to training duties with the Torpedo School H.M.C.S. STRATHCONA. She paid off into Reserve on 26.5.46 and was removed from the Canadian Pink List on 2.7.47 and was sold during 12.47 to German and Milne of Montreal who broke her up during 1948.

Note: The A to I's were transferred to Canada for a number of reasons — the Canadians already operated a number of similar vessels, so maintenance would not be a problem, the Canadians had a requirement for fast A/S vessels and it eased the manpower problems being experienced by the Royal Navy at this time.

FURY (H76)

After working up at Portland FURY joined the 6th D.F. and immediately saw service in the Mediterranean during the Abyssinian crisis of 1935/36. She then spent the remainder of her pre-war service with the Home Fleet, except for spells with the Non-intervention patrol off the Spanish Biscay and Mediterranean coasts. The 6th D.F. was re-numbered the 8th in April 1939 and nearly all her subsequent service was to be with the 8th D.F.

After spending the winter of 1939/40 screening units of the Home Fleet, based at Scapa, FURY's 'hot' war started in 4.40, when she was one of the vessels that escorted the cruiser SUFFOLK into Scapa on 17.4.40, after the latter had been badly damaged by German aircraft, after bombarding Stavanger seaplane base. On 10.5.40 FURY was one of the vessels that escorted the badly damaged KELLY into the Tyne. She was then sent south to aid the evacuation of the B.E.F. from France and continued on these duties until 7.40.

It was during this month that she and the rest of the 8th D.F. were allocated to the newly created Force 'H', based at Gibraltar. Due to defects, however, FURY did not leave the U.K. until 22.8.40 when she sailed as part of Operation 'HATS' — the reinforcement of the Mediterranean Fleet — and arrived at Gibraltar on 29.8.40.

Almost immediately, on 6.9.40, FURY joined Force 'M' — Operation 'MENACE' — the abortive invasion of Dakar. The operation was a fiasco and FURY subsequently operated off the west coast of Africa on escort duties for two months, before re-joining Force 'H'.

On 27.11.40 FURY was one of the vessels escorting the battle-cruiser RENOWN at the action off Cape Spartivento, and she later participated in the bombardment of Genoa and a flying-off operation of aircraft to Malta during 4.41. FURY had had a brief refit at Gibraltar during 3.41 and continued to operate with Force 'H' until 10.41, when she returned to the U.K. Previous to this FURY had participated in two major convoy operations to Malta — 'HALBERD' and 'EXCESS'.

On her return to the U.K. FURY operated briefly with the Special Escort Division at Greenock before rejoining the 8th D.F. of the Home Fleet. A long overdue refit was completed on the Humber on 13.2.42, and included the addition of 20mm Oerlikon guns. On completing her work up she rejoined the 8th D.F. of the Home Fleet and immediately escorted Convoy PQ13 from Iceland to Murmansk. FURY, with the cruiser TRINIDAD, damaged the German destroyer Z26. TRINIDAD was torpedoed during the action and FURY escorted her into Murmansk. FURY also escorted the ill-fated PQ17 Convoy. In 8.42 she was one of the escorts for the battle squadron that escorted the 'PEDESTAL' convoy to Malta and returned to the U.K. as escort for the battleship NELSON. In 9.42 she returned to the Arctic as one of the escorts for the PQ18 Convoy.

During 10.42 she refitted on the Humber, before resuming North Russian escort duties for the remainder of the winter of 1942/43. In 3.43 FURY was one of the fifteen Home Fleet destroyers transferred to the Western Approaches during the final U-boat crisis and was attached to the 4th Escort Group. She remained on escort duties until she refitted on the Humber during 5-6.43.

FURY, as a member of the 8th D.F., left the U.K. on 17.6.43 as a reinforcement for the Mediterranean Fleet and participated in Operation 'HUSKY' — the Sicily Landings on 10.7.43 and Operation 'AVALANCHE' — the landings at Salerno on 9.9.43. FURY then operated in the Aegean during 10-11.43 and participated in three actions, that of 7.10.43 with the cruisers PENELOPE and SIRIUS and the leader FAULKNOR when they sank an ammunition ship, a trawler and six landing craft north of the island of Stampalia. The second action occurred on 7/8.10.43 when with EXMOOR and BLENCATHRA she went close inshore looking for beached landing craft and finally, with ALDENHAM and PENN, she bombarded the island of Kos.

FURY refitted at Gibraltar between 12.43-2.44 and was also converted into an A/S destroyer losing a 4.7" gun and her 3" gun and being fitted with an additional pair of Oerlikons. She then operated with the 8th Flotilla in the Mediterranean until 5.44 when she left for the U.K. with the remainder of the Flotilla and briefly returned to the Home Fleet at Scapa on 13.5.44. She then covered the 'D' Day Landings and build-up convoys across the Channel. However, on 21.6.44 she was seriously damaged by a mine off the beachhead and was run ashore. She was later salvaged, and declared a constructive total loss and was towed to T. W. Ward's yard at Briton Ferry where she arrived on 18.9.44 for demolition.

Salvage operations seem to be in progress on FURY, beached after being mined off the assault beaches at Normandy on 21.6.44.

THE 1933 PROGRAMME
THE LEADER: GRENVILLE

There was little discussion on the design of this vessel, which followed FAULKNOR in her general arrangement. However, the D.N.C. noted that it was possible to reduce her length by 6' because improvements in the design of her boilers and engines meant that these spaces could be reduced by that margin. Shaft horse power produced, at 38,000, was the same as FAULKNOR. Her displacement was 25 tons less than that of her near sister.

This design was accepted by Board Minute 3111 of 1.11.33 with tenders being invited two days later.

The lowest tender was from Yarrow's, who proposed to fit their own side-fired boilers in the vessel. This would save some £2,000 and reduce the length of the vessel by 7' 6". Similar Yarrow boilers had just been fitted to the Jugoslav destroyer-leader DUBROVNIK. The fitting of these boilers entailed the following modifications to the Admiralty design:

(i) The galley was re-positioned, as its space was to be taken up by the funnel uptake and air supply to No. 1 boiler room.
(ii) No.3 4.7" gun was re-positioned.
(iii) Beam of the vessel increased by 9" to 34' 6".
(iv) Emergency oil and reserve feed water tanks to be removed.

Yarrow's tender was accepted to gain experience with the new type of boiler and GRENVILLE was ordered on 15.3.34 at a contract price of £275,412* made up of hull £117,760, main machinery £149,835 and auxiliary machinery £7,817.

A revised legend for GRENVILLE was approved under Board Minute 3220 of 12.7.34:

Details	FAULKNOR	11/33' G'	GRENVILLE with side-fired boilers 7/34
Length (extreme)	340'	334'	327'
Breadth	33¾'	33¾'	34½'
S.H.P.	38,000	38,000	38,000
Speed (deep)	32 knots	32 knots	32 knots
Oil fuel	490 tons	475 tons	469 tons
WEIGHTS (all tons)			
General Equipment	96	85	85
Armament	128	145	136
Machinery	550	530	539
Hull	731	705	700
Standard Displacement	1,505	1,465	1,460

*Excluding stores and armament.

THE DESTROYERS: THE GREYHOUNDS

Discussions on this group started on 16.5.33 with the intention of going out to tender on 1.11.33, with orders being placed during 3.34. This programme was generally adhered to and discussions on the design concerned the following aspects:

(i) Although the intention was to repeat the FEARLESS class, it was hoped to fit the vessels with a 5.1" gun, which was under trial. However, the 5.1" gun had an elevation of 30°, and to increase this to 40° in accordance with present policy would necessitate a new design of the mounting, which would not be ready for the 1933 destroyers. On 8.6.33, it was accepted that the 1933 destroyers would retain the 4.7" 40° mounting and this was confirmed eleven days later, when the report from the C. in C. Home Fleet and the Commodore 'D' recommended retention of the 4.7" in preference to the 5.1" gun.

(ii) Modified boiler and engine room compartments:

The length of the machinery spaces could be reduced due to more efficient and compact machinery and as a result the 1933 destroyers could be of the same length as the ACASTA class and their displacement would be appreciably less than the E/F classes. Cruising turbines were no longer to be fitted, as it was felt that wartime actions would take place at speeds in excess of 15 knots. Endurance would be 5,500 miles compared with the 6,000 miles for the E/F classes.

(iii) Incorporated in the design were the changes agreed in the newly drafted Staff Requirement for destroyers of 10.7.33:
  (a) Depth charge armament to be included.
  (b) Four destroyers per flotilla to be fitted with smoke laying apparatus.
  (c) Endurance to be increased from 4,000 to 5,500 miles at 15 knots.

The ACASTAs and subsequent vessels had been designed to meet the 1928 Staff Requirement. There was also great pressure from the sea commands to keep the size of the vessel around 1,320 tons and this had been confirmed at the Sea Lords' meeting of 19.4.29 that had discussed the CRUSADER design. This decision was confirmed by another Sea Lords' Meeting on 19.5.30.

On 1.11.33 by Minute 3112 the Board "Approved the legend and drawings of the G class destroyers of the 1933 Programme".

Details	1933 Programme	FEARLESS
Length (P.P)	312'	318¼'
Length (W.L.)	320'	326'
Breadth	33'	33¼'
Displacement (standard)	1,350 tons	1,405 tons
Draft (mean)	10' 8"	10' 10"
S.H.P.	34,000	36,000
Speed (deep)	31½ knots	31½ knots
Oil Fuel	455 tons	470 tons

Complement 137 (peace) 146 (war)
Armament for both groups of vessels comprised: four 4.7" 40°, two 0.5" machine guns, four .303" machine guns, eight 21" Torpedo Tubes, 20 depth charges.

Weights (all tons)	1933 Programme	FEARLESS
General Equipment	77	87
Armament	128	128
Machinery	490	525
Hull	655	665
Board Margin	—	—
Standard Displacement	1,350	1,405

It appears that a better vessel could have been achieved by retaining the original length of the FEARLESS, improving the crew's accommodation and increasing the endurance of the vessel with extra fuel and water. However, economy ruled the day.

Before the vessels were ordered, it was decided to fit a Thornycroft vessel with 2 quintuple torpedo tubes for evaluation. The selected vessel 'GLOWWORM' was inclined at Southampton Docks on 29.11.35 with a weight of 1,240 tons, an estimated deep displacement of 1,859 tons and a G.M. of 2'2".

On 15.6.34 it was agreed to fit Johnson boilers in GIPSY at an additional cost of £8,000. This boiler proved to be smaller, lighter and to have better durability than the Admiralty type.

Tenders were invited on 3.11.33, but the vessels were not finally ordered until 5.3.34 at the following yards:

Name	Builder	Tender Price (£)			
		Hull	Main Machinery	Aux. Machinery	Total
GREYHOUND	Vickers-	105,312	133,869	9,587	248,768
GRIFFIN	Armstrongs	105,062	133,869	9,587	248,518
GARLAND		106,515	135,438	8,711	250,664
GIPSY	Fairfield	106,215	135,438	8,711	250,364
GLOWWORM		105,860	134,218	8,707	248,785
GRAFTON	Thornycroft	105,560	134,218	8,707	248,485
GALLANT		107,200	136,122	9,598	252,920
GRENADE	Alex Stephen	106,840	136,122	9,598	252,560

* Excluding Admiralty items

PROPOSED MODIFICATIONS TO THE VESSELS' A.A. ARMAMENT

On 6.1.36 it was proposed to fit the G's (all of which had been launched by this time) with 4.7" guns with 70° of elevation. The resultant vessel would have had the following weights compared with the original specification for the GREYHOUND:

CATEGORY	GREYHOUND	NEW DESTROYER	
Displacement	1,350 tons	1,450 tons	
Dimensions	320' x 33' x 19¼' (Deep)	330' x 34' x 19½' (Deep)	
Weights (tons)			Increase
Hull	655	710	(55)
Machinery	490	510	(20)
Armament	130	160	(30)
Equipment	75	80	(5)
Standard Displacement	1,350	1,460	(110)

The proposals were not proceeded with as the vessels were too far advanced. Another proposal was made on 8.5.36 to arm the vessels with 4" twin high angle weapons. The new mounting weighed 13.65 tons compared with the 9.4 tons of that currently mounted in the GREYHOUNDs. However, the new weapons would require a gun crew of 16, necessitating an increase in the vessels' complement of approximately 30 men. The ships would have to be lengthened to provide the additional accommodation required at a weight penalty of 50 tons.

WEIGHTS (tons)	GREYHOUND	Difference	Proposed Design
Hull	655	(52)	707
Equipment	74	(8)	82
Armament	131	(32)	163
Machinery	490	(20)	510
Standard Displacement	1,350	(112)	1,462

Again the work was not proceeded with because of the advanced state of the construction of the vessels and the additional cost involved. Again the Admiralty were slow to reflect in destroyer designs the advances in air power that had occurred during the 1920s and 1930s. The need to arm destroyer type vessels with high angle weapons was long overdue as the Japanese had developed an effective high angle weapon nearly a decade before. The original destroyer design was now nearly ten years old and topweight was becoming a concern. By late 1935, the D.N.C. was saying: "If the G and H classes are accepted in their present condition, then it is considered that instructions should be issued that no alterations or additions involving increases in top weight be approved without compensating weights being landed".

There was also the need to match foreign twin-gunned vessels, especially those present in the Japanese and U.S. Navies.

1933 BUILDING PROGRAMME

NAME	BUILDER	LAID DOWN	LAUNCHED	COMPLETED
GRENVILLE	YARROW	29.9.34	15.8.35	1.7.36
GIPSY	FAIRFIELD	4.9.34	7.11.35	22.2.36
GREYHOUND	} VICKERS-	20.9.34	15.8.35	1.2.36
GRIFFIN	} ARMSTRONGS	20.9.34	15.8.35	6.3.36
GARLAND	FAIRFIELD	22.8.34	24.10.35	3.3.36
GALLANT	} ALEX	15.9.34	26.9.35	25.2.36
GRENADE	} STEPHEN	3.10.34	12.11.35	28.3.36
GRAFTON	} THORNYCROFT	30.8.34	18.9.35	20.3.36
GLOWWORM		15.8.34	27.7.35	22.1.36

GRENVILLE (HO3)

Commissioned in 7.36, GRENVILLE immediately joined the Mediterranean Fleet for a brief period with the 20th D.F. and then as leader of the First D.F. She spent the first ten months in the Western Mediterranean off the Spanish coast before returning to Portsmouth for repairs between 24.5-9.6.37. She then returned to the Mediterranean until 5.38.

Refitted at Portsmouth between 7.6-25.7.38, she again returned to the First D.F. in the Mediterranean until 10.39.

GRENVILLE about to dress overall with masthead ensigns. The picture gives a good indication of the extended superstructure and Q gun positioned between the funnels.

On 22.10.39, she left Gibraltar with GIPSY, GRENADE and GRIFFIN and arrived at Plymouth on 2.11.39. GRENVILLE participated in unsuccessful submarine hunts off Start Point on 2.11.39 and then off Ushant three days later.

On the night of 7/8.11.39, GRENVILLE and GRENADE were in collision in Plymouth Harbour, GRENVILLE being holed below the waterline with No. 3 boiler room being flooded. Repairs undertaken at Devonport were not completed until 1.12.39.

Three days later GRENVILLE arrived at Harwich, still as leader of the First Destroyer Flotilla, which had been re-allocated to the Nore Command. She then operated on patrol and escort duties for the next two months. The severe weather being experienced had forced the Germans to convoy materials from Holland to Germany by sea rather than use the canals and rivers that were frozen. GRENVILLE participated in two sweeps — operations ST1 and ST3 on 15-16 and 18-19.1.40 to try to halt this traffic.

Whilst returning from the latter operation at 12.50 hours on 19.1.40 and when 23 miles east of the Kentish Knock light vessel in position 51°39'N 02°17'E, GRENVILLE was mined and sank with the loss of 4 officers and 73 ratings.

GALLANT (H59)

The third of the class to complete, on 25.2.36, GALLANT joined the First D.F. of the Mediterranean Fleet for all her pre-war service. She returned to the U.K. to refit at Sheerness between 31.5-21.7.37. She was to spend much time off the Spanish Mediterranean coast and on 20.12.36 she towed off a Spanish vessel which had run aground between Almeria and Malaga. On 6.4.37 she was bombed off Cape San Antonio by a Nationalist aircraft, but not hit, whilst on 'Non-Intervention' duties.

GALLANT suffered a lingering end after being mined off Malta on 7.1.41. Whilst under repair she suffered further damage and was finally declared a constructive total loss.

On 22.10.39 GALLANT berthed at Plymouth following the transfer of the Flotilla to the Western Approaches Command. After boiler cleaning at Plymouth, she participated in the search for a German merchant ship that had been reported as having left Vigo, but without success.

Five days later the First D.F. including GALLANT transferred to the Nore Command at Harwich for patrol and escort duties until 1.2.40 when, with her sister GRIFFIN, she was detached to Rosyth on A/S duties.

On 2.2.40 GALLANT and GRIFFIN stood by the tanker BRITISH COUNCILLOR mined in position 53° 48' N 00° 34' E whilst part of convoy FS84 and returned to Rosyth the next day with the vessel's crew. On 18.2.40 she was detailed to escort convoy HN12 following the loss of its escort — the destroyer DARING. A week later GALLANT picked up 12 survivors from the Swedish vessel SANTOS torpedoed and sunk 50 miles east of Duncansby Head the previous evening. On 20.3.40 she provided A/S protection for the A.M.C.s CILICIA and CARINTHIA in collision that day in position 59° 17' N, 00° 42' W.

GALLANT then refitted at Southampton between 28.3-30.4.40 rejoining the First D.F. at Harwich the next day. On the night of 9/10.5.40 GALLANT and BULLDOG took off most of the crew of the destroyer KELLY after she had been torpedoed by an S-boat in the North Sea (see BULLDOG).

GALLANT then participated in the Dunkirk evacuation and had already taken off some 1,880 troops when at 11.55 hours on 29.5.40 she was attacked by seventeen dive bombers off the Snouw Bank, when in company with the destroyers GRENADE and JAGUAR. GALLANT was near missed by a bomb that exploded 10 yards from her stern, which flooded her tiller flat, knocked out her steering and caused minor structural and electrical damage. She was then withdrawn from 'DYNAMO' and repaired at Hull, but was operational again by the night of 5/6.6.40, when with the destroyer WALPOLE she encountered a German minelaying sortie off Lowestoft. GALLANT also participated in an unsuccessful attempt to intercept the German battle-cruiser SCHARNHORST at large in northern waters on 21-22.6.40.

After repairs which included the fitting of a 3" H.A. gun, at Chatham, GALLANT with GREYHOUND made passage for the Clyde, where they joined the escort for the aircraft carrier ARGUS and A.M.C. MALOJA which left the Clyde on 24.7.40 for Gibraltar. Off Northern Ireland the vessels rendezvoused with the destroyers ENCOUNTER and HOTSPUR escorting the REINA DEL PACIFICO and CLAN FERGUSON carrying supplies for the Mediterranean via the Cape. GALLANT arrived at Gibraltar on 30.7.40 and joined the 13th D.F. of the North Atlantic Command at Gibraltar. She then participated in the convoy operation 'HATS' with Force 'H' before returning to Gibraltar on 5.9.40.

GALLANT operated with Force 'H' for the next two months, sinking with HOTSPUR the Italian submarine LAFOLE south east of Alboran Island in position 35° 50' N 02° 53' W on 20.10.40. Nine survivors were captured. She also shadowed French destroyers that were proceeding to Casablanca and supported convoy HX84 after it had been attacked on 5.11.40.

She joined the Mediterranean Fleet as part of Force 'F' at 10.15 hours on 10.11.40 and entered Valetta Harbour two hours later. On 23.11.40 with GRIFFIN and GREYHOUND she joined the 14th D.F. On 27.11.40 whilst participating in the reinforcement Operation 'COLLAR' she fought in the indecisive action off Spartivento.

On 7.1.41 GALLANT left Malta as escort for the Malta Convoy — Operation 'EXCESS'. At 08.30 hours on 10.1.41 some 25 miles south-east of Pantellaria in position 36° 24' N 12° 10' E GALLANT activated a mine.

The explosion occurred under 'A' mounting and the fore part of the vessel disappeared. It seems probable that the foremost magazine ignited and that an explosion was only prevented by the inrush of water. The bows floated away with the stem horizontal and 20' out of the water. One man escaped from this portion of the ship. The bridge structure was intact but the fore part was missing as far back as No. 1 boiler room, approximately at No. 40 station. No. 60 bulkhead was strained and leaking and steam was let off as it was felt that the ship might sink. No. 1 boiler room was flooded. 65 ratings, but no officers, were killed or missing, and another 15 ratings were injured. Her sister GRIFFIN embarked the wounded and the majority of the survivors.

GALLANT was towed stern first into Malta by the destroyer MOHAWK and arrived at Malta at 11.30 hours on 11.1.41. She paid off into dockyard control and repairs continued until 4.42. During 10.41 it was estimated that her repairs would be completed in 6.42.

However, on 5.4.42 she suffered extensive splinter damage from near misses during an air raid on Valetta and had to be beached at Pinto Wharf. Subsequently, she was declared a constructive total loss and stripped of useable equipment.

In 9.43 the hulk of GALLANT was utilised as a blockship at St. Paul's Island, Malta. The wreck is reported to have been dispersed during 1953.

GARLAND (O.R.P. GARLAND) (H37)

GARLAND operated with the First D.F. of the Mediterranean Fleet from commissioning until the outbreak of war. She refitted at Sheerness between 24.5-5.7.37 and 31.5-28.7.38 during which repairs were also made to her low pressure turbines. She operated off southern Spain between 11-12.38 and 4-5.39 and Cyprus during 7.39.

On 3.9.39 she was on passage from Aden to Alexandria and finally arrived at Alexandria 6.9.39. On 17.9.39 whilst on convoy escort duty between Alexandria and Malta she suffered a premature explosion of her own depth charges. She suffered extreme buckling of the upper decks from the after engine room bulkhead to the stern. She was towed into Alexandria where surveys revealed that extensive renewal of plates and structures were necessary as were repairs to her machinery. After temporary repairs had been undertaken, the ship had to be towed to Malta, where she arrived on 10.10.39. She was in hand for repairs at Malta between 11.10.39-8.5.40.

GARLAND was turned over to the Polish Navy on 3.5.40 — the anniversary of Poland's Constitution — and operated in the Mediterranean as an unallocated vessel on local patrol duties between Alexandria and Malta. On 31.8.40 GARLAND and the transport CORNWALL were slightly damaged by Italian air attack during the passage of the reinforcement convoy 'HATS'. On 14.9.40 GARLAND left Gibraltar on passage to Plymouth to join the Western Approaches Command's 10th Escort Group until 4.41.

On 13.11.40 GARLAND and vessels of the 3rd D.F. sailed as an escort for the battleship REVENGE from the Clyde to 20° West in gale force conditions which made it impossible for GARLAND to keep up. GARLAND suffered heavy damage to her superstructure and two men were lost overboard. Repairs at Harland and Wolff's Govan Yard were completed between 17.11-26.12.40. She then resumed her escort duties until 7.41 after being fitted with a new ASDIC set at Barclay Curle's in Glasgow between 4-14.1.41. In 4.41 she joined the 14th Escort Group based at Greenock.

During 7.41 she was attached to the Home Fleet and during the Spitzbergen Raid she escorted a tanker supporting the fleet's operations. On her return to Western Approaches Command she joined Escort Group B3 until 5.42. Between 24-30.9.41 GARLAND and PIORUN escorted the 'HALBERD' convoy from Gibraltar to Malta. On the latter date, both vessels arrived at Gibraltar escorting the damaged battleship NELSON which had been torpedoed during the convoy action. GARLAND then resumed her escort duties with B3 Group and was embroiled in Arctic convoy duties to Kola and Archangel. She was refitted at Smith's Dock

GARLAND in 5.42 whilst operated by the Polish Naval Service. A few days later, on 27.5.42, she suffered considerable damage and casualties during an air attack off Norway.

Middlesbrough between 28.2-5.5.42 and after a brief work up joined the escort for Convoy PQ16 for Seidisfjord on 21.5.42. Six days later on 27.5.42 on the third day of determined air raids GARLAND was attacked by seven Ju88's at 13.55 hours. The aircraft approached from the starboard beam and six dropped bombs on vessels of the convoy, the other aimed its bombs at GARLAND.

The bombs fell 10 yards to the starboard side of the ship in line with B gun. 26 of her crew were killed, one was missing and 37 wounded, some of whom subsequently died, when havoc was caused to the personnel manning A and B guns, as well as the starboard Oerlikon and machine gun. Range-finder and fire control equipment was irreparably damaged and A and B guns were left without crews. The vessel suffered some 500 splinter holes. Although there was widespread damage to the crew's quarters, boats and deck equipment, GARLAND suffered very little underwater damage.

GARLAND was detached to run at maximum speed to Murmansk, where she arrived on 29.5.42. The ship was under repair there for the whole of 6.42 and finally left on 4.7.42 and arrived at Troon where repairs were completed between 8.7-21.9.42. On completion of her work up GARLAND rejoined Escort Group B3 on Atlantic Escort duties until 5.43 based at Greenock. Between 1-16.2.43 she repaired minor defects and repairs to her range-finder at Greenock.

Between 1-5.43 GARLAND participated in the momentous convoy actions that were being fought against the U-boats in the North Atlantic. GARLAND with the frigate SWALE and the corvettes NARCISSUS and ORCHIS protected Convoy SC117 from 22.1.43 and HX228 between 10-12.3.43. A prolonged refit then followed on the Clyde between 27.5-8.9.43.

She then joined the 8th Escort Group between 9-11.43 on the North Atlantic run, and her duties included escorting aircraft carriers to the Azores during 10.43. A period to rectify defects then followed on the Clyde 5.11-2.12.43, before O.R.P. GARLAND made passage to Freetown to undertake escort duties between Freetown and Gibraltar until 4.44. She was then transferred to the 14th D.F. of the Mediterranean Fleet until 11.44. During this period in warmer waters, GARLAND undertook A/S sweeps and aided in the occupation of several Greek islands. On 18.9.44 whilst operating off Santorini in the Aegean she attacked U407 which after an extended hunt was finally sunk by Hedgehog and depth charge attack early the next morning. Repairs at Alexandria to damage caused by depth charge explosions in these actions took 10 days. Further gun support duties followed at Kassandra and in the Cyclades. On 20.11.44 GARLAND departed from the Mediterranean for a refit at Devonport, which was not completed until 31.3.45. She had barely worked up as a vessel of the 8th D.F. at Plymouth when the war ended. After being employed on the carriage of emergency supplies to the populations of Dutch and Belgian towns GARLAND operated in Home Waters.

During 11-12.45 GARLAND acted as one of the escorts for the scuttling of German U-boats in the Atlantic during Operation 'DEADLIGHT' and then operated on miscellaneous escort duties in Norwegian waters between 1-3.46. She had been a unit of the 23rd D.F. of the Home Fleet between 6-9.45 and then the 17th Flotilla until 3.46. Between 3-6.46 she formed part of the Polish Squadron at Rosyth.

On 23.7.46 it was announced she was to be paid off and the loan terminated. Between 19-28.8.46 her armament was removed before she entered Category 'C' Reserve at Harwich. However, she was to have a reprieve. On 14.11.47 she was sold 'as lies' to the Dutch for £9,000. She initially operated at Amsterdam as a floating tender for technical training in the Royal Netherlands Navy. However, she was soon taken to the yard of the Netherlands Dock and Shipbuilding Company during 1948 and converted into an A/S training vessel.

MARNIX (ex GARLAND) showing the considerable modifications made to fit her for her duties as a training vessel in the Royal Netherlands Navy. (Royal Netherlands Navy)

On 16.1.50 she was renamed MARNIX and made a training cruise to the south coast of Britain during 3.50 with the submarine O24. In 1952 MARNIX was re-classified as a frigate and continued with her duties for the next five years. During 1955-56 she underwent considerable repairs at Willemstad and continued in service until 31.1.64 when she was finally decommissioned after nearly 30 active years.

GIPSY (H63)

GIPSY briefly joined the 20th D.F. of the Home Fleet, before joining the Mediterranean Fleet's First D.F. from 3.36 until 8.38. She then returned to the U.K. to refit at Devonport between 2.6-30.7.38.

GIPSY'S was career was to be all too brief. She was mined off Harwich on 21.11.39, a few weeks after her return from the Mediterranean. (W.S.P.L.)

She then returned to the Mediterranean and the First D.F. until recalled to the U.K. during 10.39. On 11.11.39 she arrived at Plymouth on passage to join the 22nd D.F. at Harwich. The next day she was slightly damaged in a collision with her sister GREYHOUND, whilst on passage to Harwich, where she arrived the same day. She repaired her damage and was made ready for service, but was to survive a mere nine days.

On 21.11.39 she rescued three German airmen from the channel outside Harwich and returned with them to Parkeston Quay where they were handed over to the army escort.

At 21.00 hours on the same day, GIPSY, with GRIFFIN (Captain D22), KEITH and BOADICEA weighed anchor and started to go out on patrol in the North Sea. At 21.20 hours, as she was leaving harbour, she activated a magnetic mine, which exploded amidships under the engine room. (The mine had probably been laid by the aircraft, the crew of which had been rescued that afternoon.)

The other ships of the flotilla came alongside and started to pick up survivors and wounded and landed them at Parkeston Quay. GIPSY's C.O. Lt. Cdr. N. J. Crossley who had been thrown from the bridge and had landed on the forecastle died of his wounds on 27.11.39. 29 other members of the crew were killed, but 115 survived.

The vessel was lying in two pieces held together by buckled plates and a wrecked gun platform between her funnels. The ship was seen to be in an upright position, down on the seabed with only the bridge superstructure visible at high tide, but blocking the fairway. Clearance operations began immediately. The two halves of the vessel were at first severed by explosives and both parts were raised separately by the use of camels.

The wreck was bought by the Westminster Dredging Co. Ltd. of London, which resold the two halves to Stow Salvage Ltd. which finally broke up both halves for scrap. T. W. Ward's records show that 750 tons of ferrous scrap and 38 tons of non-ferrous metals were recovered between 6.40-2.44.

GLOWWORM (H92)

After working-up at Portland during 2.36, GLOWWORM left for the Mediterranean where she joined the First D.F. until 5.37. Her duties were similar to those of her leader GRENVILLE. After giving leave, whilst repairs were completed at Portsmouth between 27.5-8.6.37, she returned to the First Flotilla until 5.38, when she returned to the U.K. to refit. Refitted at Portsmouth between 7.6-25.7.38 she again returned to the Mediterranean. At the time of the Munich crisis she escorted the P&O liner STRATHNAVER from Malta to Alexandria between 20-22.9.38 and the cruiser ARETHUSA to Aden between 22-26.9.38.

GLOWWORM dressed overall for the Coronation Review of 1937.

The crisis over, she returned to her normal peace-time duties until 16.5.39 when she was in collision with GRENADE during night exercises. Badly holed forward, she proceeded to Alexandria, escorted by GRENVILLE, returning to Malta on 22.5.39. Damage repairs were completed there between 23.5-24.6.39.

On 3.9.39 she was at Alexandria and operated in the Mediterranean until Italy's intentions became clear, GLOWWORM finally left Gibraltar for Home Waters on 19.10.39, arriving at Plymouth three days later to join the Western Approaches Command. This period of service was brief as on 12.11.39 she arrived at Harwich on attachment to the 22nd D.F. of Nore Command.

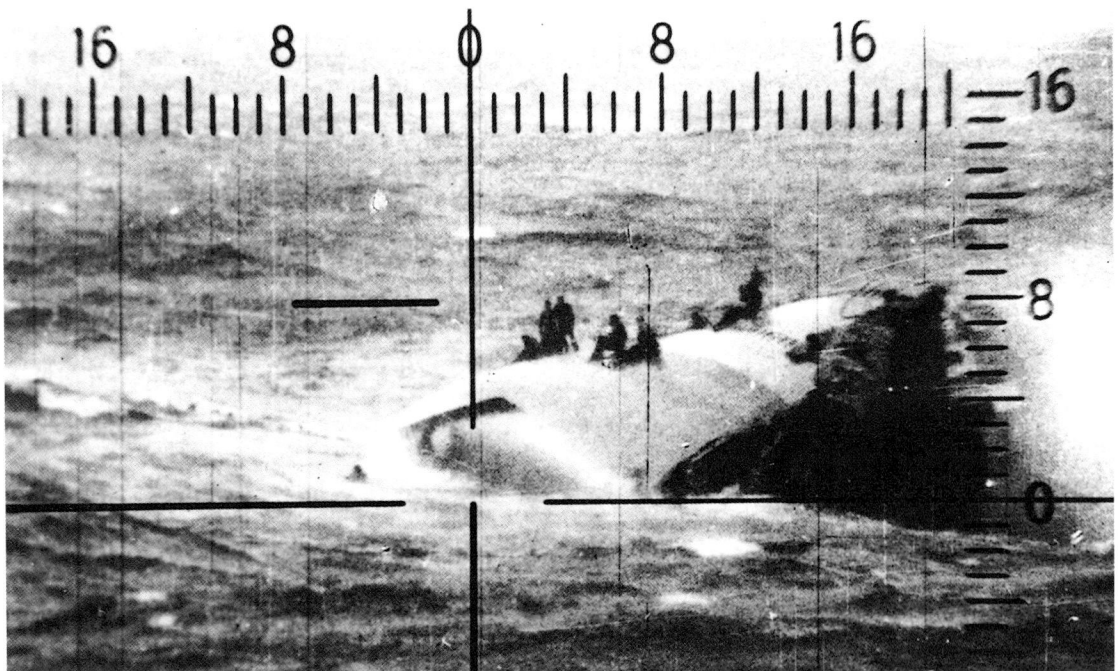

GLOWWORM'S final moments as seen through the gunsights of the German cruiser ADMIRAL HIPPER, which she had rammed off the Norwegian Coast on 8.4.40. Her Commander, Lt. Cmdr. Roope, was awarded a posthumous Victoria Cross.
(Imperial War Museum HU2212)

On 22.2.40 whilst at anchor she was hit by the Swedish ship REX in dense fog, off the Outer Dowsing and was considerably damaged. Between 26.2-20.3.40 she was at the yard of Brigham and Cowan at Hull for collision repairs. On her return to the First Flotilla, she was attached to the Home Fleet at the start of the Norwegian Campaign.

She left Scapa on 5.4.40 to intercept ships reported by aircraft. However, she had to drop back to search for a man overboard and lost contact with the battle-cruiser RENOWN as she approached the Norwegian coast. At about 08.30 hours on 8.4.40 she reported that she was in contact with two German destroyers some 140 miles from RENOWN. Further signals indicated that she was engaging a superior force and her last signal was timed at 08.55 hours.

It transpired that GLOWWORM had engaged the German heavy cruiser ADMIRAL HIPPER attacking that vessel with torpedoes and although considerably damaged by enemy fire, she rammed the German ship. She sank soon afterwards in position 64°27′ N 06°28′ E. Her C.O., Lt. Commander G.B. Roope. six other officers and 105 ratings were lost with one officer and 39 ratings surviving to become prisoners of war.

In 1945, when the details of this gallant action were made known, her C.O. was awarded a posthumous Victoria Cross.

GRAFTON was the first of two 'G' class destroyers lost on 29.5.40 at the height of the Dunkirk evacuation. She was torpedoed by U62.

GRAFTON (H89)

On completion on 20.3.36, GRAFTON worked up and briefly joined the 20th D.F. before transferring to the First D.F. in the Mediterranean during 6.36. She was to serve with this Flotilla until the outbreak of war. However, between 10.8-9.9.36 she escorted the yacht NAHLIN, with H.M. King Edward VIII aboard, cruising in the eastern Mediterranean. GRAFTON then operated extensively off the Spanish coast, evacuating refugees in the opening weeks of the Civil War and then undertaking Neutrality patrols.

In 9.39 GRAFTON was under repair at Malta and as soon as these repairs were completed she was transferred to the Western Approaches Command with her sisters GALLANT, GLOWWORM and GREYHOUND and arrived at Plymouth on 22.10.39. After undertaking patrol and escort duties for a few weeks she was transferred to Harwich in the following month as a replacement for KEITH in the 22nd D.F. However, on 10.1.40 GRAFTON with the rest of her division, joined three Polish and two other destroyers to form a re-constituted First D.F. still based at Harwich.

The destroyers were employed on the interception and examination of vessels sailing between German and Dutch ports. On 16.1.40, GRAFTON brought the Latvian vessel RASMA into Dover for examination after she had been intercepted off the Maas Light Vessel. On 11.2.40 a party from GRAFTON boarded the Dutch vessel GRADA but the vessel was released.

After a short refit at the yard of Brigham and Cowan at Hull between 26.3-14.4.40 GRAFTON was attached to the Home Fleet. She operated in Norwegian waters until 11.5.40, when she left for Scapa. She returned to the Nore Command on 24.5.40 to participate in Operation 'DYNAMO' — the evacuation of Dunkirk.

On 26.5.40 she escorted the cruisers ARETHUSA and GALATEA during their bombardment of enemy positions at Calais and the next day she finally joined Operation 'DYNAMO'. On 28.5.40 she evacuated 860 troops from the beaches of La Panne and Braye and disembarked them at Dover in the late afternoon. GRAFTON immediately returned to Dunkirk and again picked up over 800 men off the Braye dunes and set off for Dover via the Kwente Buoy to the north-east of Dunkirk.

At 03.09 hours the next day (the 29th), in the vicinity of Kwente buoy, GRAFTON was torpedoed by U62. She had been in the act of rescuing survivors from the destroyer WAKEFUL, torpedoed earlier by the German S62.

The torpedo struck the destroyer on the port quarter forward of the propeller, causing severe structural damage aft, but her main machinery and lighting were still in operation. A second explosion then wrecked the whole fore part of the bridge, killing her C.O. Commander G. E. C. Robinson and another officer. GRAFTON, however, remained afloat, immobilised with her back broken. At 04.00 hours the transport MALINES and the destroyer IVANHOE took off all the troops and her crew. IVANHOE then sank the GRAFTON by gunfire. Four lives were lost — a remarkably low figure in the circumstances. The wrecks of GRAFTON and WAKEFUL lie together in position 51° 22' 14" N 02° 39' 57" E.

GRENADE (H86)

After completion on 28.3.36, GRENADE completed her trials and work-up before joining the First D.F. of the Mediterranean Fleet the following month until the outbreak of war. Based at Malta, she refitted there between 20.3-24.4.37, but returned home to give leave and refit at Chatham between 27.5.38 and 7.38. A brief detachment to the Red Sea followed during 10.38. GRENADE was lying at Alexandria when war was declared.

GRENADE was the second vessel of the class lost on 29.5.40. She succumbed to air attack between Dover and Dunkirk.

After undertaking patrol duties in the Mediterranean, she left the Mediterranean station on 25.10 39 and arrived eight days later at Plymouth where she joined her flotilla as part of the Western Approaches Command. She was to remain with the First Flotilla until her loss. On 7.11.39, whilst at Devonport she collided with her leader GRENVILLE and had her stem twisted and her fore-peak flooded. Repairs at Falmouth were completed on 9.12.39 and she rejoined her flotilla, now based at Harwich.

For the remainder of 12.39-1.40, GRENADE was engaged in contraband control work in the Downs. After such a search, GRENADE and GRIFFIN picked up 15 officers and 102 ratings of GRENVILLE 23 miles east of the Kentish Knock light vessel, after the leader had been mined and sunk. GRENADE returned to Harwich with the survivors at 02.00 hours on 20.1.40.

GRENADE then refitted at London between 27.1-27.2.40, but was prevented from sailing after completion of the refit by a collision with R.M.S. ORION that same day, while the latter was berthing. After temporary repairs had been completed, GRENADE arrived at Harwich 2.3.40 for permanent repairs to the approximately 20ft of her hull which was indented. She re-entered service on 3.4.40 and was attached to the Home Fleet at Scapa.

Immediately GRENADE and ENCOUNTER left Scapa as escort for the oiler BRITISH LADY which arrived safely at Skjaelfjord on 12.4.40. Two days later GRENADE and ENCOUNTER joined the battleship WARSPITE off Narvik and later provided the escort for the aircraft carrier ARK ROYAL off Namsos and Andalsnes. GRENADE did not arrive back at Scapa until 28.4.40.

On 3.5.40, GRENADE and IMPERIAL stood by the French destroyer BISON which had been hit and damaged by aircraft bombs. GRENADE subsequently sank the BISON after the latter's forward magazine exploded and rescued her survivors before returning to Scapa on 5.5.40.

GRENADE was then ordered to the Channel and arrived at Dover on 13.5.40. The next day, whilst fog-bound in position 51° 05′ N 01° 21′ E, she collided with the anti-submarine trawler CLAYTON WYKE and was holed above the waterline. Repairs were completed at Sheerness on 25.5.40.

For the first days of the 'DYNAMO' operation GRENADE, with the destroyers CODRINGTON, JAGUAR and JAVELIN acted as a covering force in the Brown Ridge Buoy area and later established a patrol line between the West Hinder and Kwente buoys. On 28.5.40, GRENADE with ANTHONY, CODRINGTON and JAVELIN picked up 33 survivors from the s.s. ABUKIR, which had been torpedoed by an S-boat.

After making one return trip to Dunkirk on 28/29.5.40, GRENADE with GALLANT and JAGUAR left Dover on 29.5.40 for Dunkirk. GRENADE survived two bombing attacks, but at 18.00 hours whilst in Dunkirk harbour, she was attacked for a third time and was hit by three bombs in quick succession. The first hit occurred on the starboard side between A and B 4.7" guns, causing a fire at the after-end of the fore mess deck and making a hole in the side plating. The second bomb hit near the sick bay and penetrated almost to the bottom of the ship before exploding, whilst the third hit 'X' 4.7" gun shield and bounced into the water. An attempt was made to tow her from the fairway, but this had to be abandoned by the crew as GRENADE was on fire and she sank shortly afterwards.

GREYHOUND achieved fame at the Battle of Matapan, when she illuminated the Italian Fleet with her searchlight. She was lost two months later off Crete.

GREYHOUND (H05)

After working up at Portland during 2.36 and a brief period of service with the 20th D.F., GREYHOUND joined the First D.F. in the Mediterranean for the remainder of her pre-war service. She refitted at Portsmouth between 7.6-23.7.38 and during the Munich crisis she escorted the liner STRATHNAVER between Malta and Alexandria 22-26.9.38 and then joined the escort for the cruiser ARETHUSA to Aden.

On 19.10.39, she left Gibraltar for the U.K. and arrived at Plymouth four days later. She was briefly attached to the Western Approaches Command and did not finally leave for Harwich and 22nd D.F. with GIPSY until 12.11.39. These vessels were in collision in fog whilst on passage, but with little damage. Collision repairs were completed at Sheerness in two days. She was soon active again rescuing survivors from the s.s. SIMON BOLIVAR mined on 18.11.39 and TORCHBEARER, the next day.

On 5.12.39 she rejoined the First D.F. still based at Harwich, giving protection to convoys. On 11.2.40 she lost a midshipman and six ratings when a boat capsized as they were about to board a ship for investigation of her cargo. She refitted at the yard of Amos and Smith between 16.2-18.3.40 and was at Scapa with the Home Fleet on 4.4.40. The next day, she left Scapa with GLOWWORM as escort for the battle-cruiser RENOWN giving cover to destroyers of the 20th Minelaying Flotilla which were laying a field off Hovden in the Vestfjord. GREYHOUND was present when RENOWN was in action with the German battle-cruisers GNEISENAU and SCHARNHORST and GNEISENAU was hit twice by RENOWN before contact was lost.

The next day, GREYHOUND covered the withdrawal of the three survivors of the 2nd D.F. after the action at Narvik and did not arrive back at Scapa until 17.4.40. However, the following day, she was damaged in a bombing attack and needed repair. She left Scapa on 19.4.40 and arrived at the Gravesend yard of Green & Silley Weir on 22.4.40. Repairs were not completed until 19.5.40.

GREYHOUND then rejoined the First D.F. at Dover until 7.40. On 25.5.40, with GRAFTON she participated in bombardment duties off Calais, until the city fell. She was then involved in the Dunkirk evacuation. On 28-29.5.40 she evacuated 1,360 men before being damaged by air attack, with casualties to her crew and evacuated soldiers. She was taken in tow by the Polish destroyer BLYSKAWICA and was finally assisted into Dover by tugs. Repairs were completed at Chatham on 17.6.40. She returned to her duties at Dover, before being ordered to the Clyde to escort the aircraft carrier ARGUS with GALLANT to Gibraltar. On 30.7.40 she joined the 13th D.F. until 11.40.

Force 'H' led an active existence and GREYHOUND participated in Operation 'HATS' — the passing of reinforcements through the Mediterranean on 30.8.40 and Operation "MENACE". Whilst engaged in this operation she engaged the French destroyer L'AUDACIEUX on 23.9.40 with the Australian cruiser AUSTRALIA and FURY. The French destroyer was beached and burnt out. GREYHOUND returned to Gibraltar on 15.10.40.

On 7.11.40 GREYHOUND left Gibraltar for Operation 'COAT' — the transport of troops to Malta and the reinforcement of the Mediterranean Fleet. The troops were disembarked at Malta three days later and GREYHOUND went on to Alexandria to join the 14th D.F. until her loss. She participated in the Cape Spartivento action later the same month.

GREYHOUND spent the first months of 1941 on escort duties, especially to Piraeus. On 6.3.41 whilst escorting Convoy AS16 in the Aegean, GREYHOUND attacked and damaged the Italian submarine ANFITRITE by depth charges. She then finished the submarine off by gunfire and rescued her crew of 43. Three weeks later, GREYHOUND participated in the Cape Matapan action and started the action by illuminating an enemy cruiser with her searchlight.

4.41 saw GREYHOUND operating with the Australian destroyer VENDETTA off the North African coast carrying out sweeps against enemy shipping, sinking the sailing vessel ROMAGNA off Apollonia on 17.4.41.

In 5.41, after escorting a convoy to Malta, GREYHOUND participated in the evacuation of Commonwealth troops from Crete and patrols searching for German surface invasion fleets. On 22.5.41 she sank a caique in the Anti-Kithera Channel and whilst returning to the British Fleet she was attacked by eight Ju 87 dive-bombers and hit three times. GREYHOUND sank stern first, 15 minutes after the first attack. Her survivors were picked up by the destroyers KANDAHAR and KINGSTON, but her C.O., Commander W.A Marshall-Adeane, five other officers and 74 ratings were killed and 4 ratings became prisoners of war.

GRIFFIN/H.M.C.S. OTTAWA (II) (H31)

On commissioning GRIFFIN joined her sisters on the Mediterranean Station as a member of the First D.F. She was to remain there until 11.39 when she was recalled to Home Waters. The highlight of her Mediterranean service was the search for the KRONPRINS OLAV, drifting without radio, with her shaft broken. The search was initiated after four persons from the vessel had been picked up by the R.F.A. OLYNTHUS and taken to Malta. The netlayer PROTECTOR and the destroyers GREYHOUND and GRIFFIN searched for the vessel, which was located by GRIFFIN and brought safely into Malta. During the Munich crisis GRIFFIN escorted the P&O liner STRATHNAVER in the Eastern Mediterranean between 22-23.9.38 and later escorted the cruiser ARETHUSA from Alexandria to Aden between 26-30.9.38.

On 2.2.39 GRIFFIN collided with the target destroyer SHIKARI and was holed above the waterline near the stern. Repairs were completed by 7.2.39. On the outbreak of war she was lying at Alexandria.

On 9.11.39, GRIFFIN arrived at Plymouth from the Mediterranean and a week later joined the First D.F. at Harwich until 4.40. Almost immediately she was holed whilst at Harwich and was under repair at Harland and Wolff's East India Docks repair depot until 6.12.39. GRIFFIN then operated on normal patrol and escort duties until transferred to the Home Fleet during 4.40 for service in Norwegian waters for the next two months. On 24.4.40 she captured a German trawler in the North Sea which was carrying U-boat supplies. Later she participated in the evacuation of Namsos.

A fine view of GRIFFIN shortly after commissioning. She served in all theatres of operations, before refitting as an A/S destroyer and becoming the Canadian OTTAWA (II) as illustrated on page 6.

GRIFFIN then transferred to the 13th D.F. at Gibraltar and operated with Force 'H' between 8-11.40. On 20.10.40 GRIFFIN with GALLANT and HOTSPUR sank the Italian submarine LAFOLE off Gibraltar.

In 11.40 she was one of a batch of reinforcements transferred to the Mediterranean Fleet and joined the 14th D.F. until 12.41. She deployed to the Red Sea to complete the capture of Italian Forces in East Africa during 2-3.41.

On 7.1.41 she picked up survivors and wounded from GALLANT when the latter was mined off Pantellaria. GRIFFIN was a participant in the Cape Matapan action, before being embroiled in the Greek and Cretan campaigns during which she evacuated 720 troops from Suda Bay. She seems to have suffered no damage in either campaign and successfully completed four months' duty on the Tobruk escort run between 7-11.41. In 12.41 she joined the 2nd D.F. but continued to serve in the Mediterranean until 2.42.

In 2.42 the 2nd D.F. was sent as a reinforcement to the newly established Eastern Fleet based at Kilindini in Kenya. GRIFFIN was engaged on convoy escort duties between East Africa and South Africa until 9.42. She was borrowed by the Mediterranean Fleet for the unsuccessful 'VIGOROUS' Convoy operation to Malta from Alexandria during 6.42. Again she was a lucky ship, as there are no reports of her being damaged or suffering a casualty.

However, by 9.42 GRIFFIN was in desperate need of a refit and she returned to the UK during 10.42 and immediately started a refit at Thornycroft's Southampton yard and was at the same time converted into an escort destroyer between 2.11.42-24.3.43. On 1.3.43 she was taken over by the Royal Canadian Navy and on 20.3.43 she was commissioned by the Canadians. However, on 10.4.43 despite the objections of her Captain who wished to retain her name she was renamed OTTAWA (II) to commemorate the destroyer lost some six months earlier and to fit into the River nomenclature used by the Canadians for their destroyers. On 15.6.43 she was given as a gift to the Canadians.

OTTAWA after working up at Tobermory operated as Senior Officer with Escort Group C5 as a mid ocean escort, based at St. John's, Newfoundland until 5.44. During that month she transferred to the 11th Escort Group, as S.O. for invasion duties and during post-invasion patrols, with KOOTENAY, sank no fewer than 3 U-boats. The first was U678 on 6.7.44 off Beachy Head with the British frigate STATICE. On 16.8.44 with CHAUDIERE they sank U621 off La Rochelle. U984 was the final U-boat sunk, with CHAUDIERE west of Brest two days later.

On her return to Canada OTTAWA refitted at St. John's between 12.10.44-26.2.45. However, on 11.3.45 whilst on an anti-submarine sweep off Halifax, she was in collision with the corvette STRATFORD. Both vessels received considerable bow damage. Repairs were not completed until 30.4.45. OTTAWA then undertook several trooping voyages between St. John's, Halifax and Greenock. She was finally paid off on 31.10.45 at Sydney and was sold during 8.46 to the International Iron and Metal Company and broken up.

THE 1934 PROGRAMME

THE LEADER: HARDY

There was little or no discussion over the design of the vessel. Board Minute 3219 of 12.7.34 stated:

"The Board approved the legend of the 'H' class leader of the 1934 Programme, which was to be built as a repeat of the original 'G' class leader of the 1933 Programme. The legend and drawings of which had been approved by Board Minute 3111".

There were in fact some minor differences, as HARDY's breadth was increased by 3" over GRENVILLE, but displacement was reduced by 10 tons because the armament weighed 9 tons less and the hull one ton.

Tenders to construct the vessel were invited on 15.8.34. The tender, dated 25.9.34 from Cammell Laird's was accepted in the sum of £278,482*, made up of the following:

Hull	£120,855
Main and Auxiliary Machinery	£149,532
Other Machinery	£8,095

HARDY was formally ordered on 12.12.34.

* Plus Armament and Equipment supplied by the Admiralty.

THE DESTROYERS: THE HEROS

Originally it had been intended to fit some or all the 1934 Programme destroyers as minelayers. However, the trials of the 2 E's as minelayers were not scheduled until 11.34 and to wait for the results would have imposed an unacceptable delay on the 1934 Programme vessels.

Under Board Minute 3218 of 12.7.34 approval was given "to the legend of the 'H' class destroyers of the 1934 Programme, which were to be repeats of the 'G' class destroyers of the 1933 Programme. The legend and drawings had previously been approved by Board Minute 3112".

The weights of proposed vessels were:

Machinery	490 tons
General Equipment	77 tons
Armament	128 tons
Hull	655 tons
Standard Displacement	1,350 tons

Tenders for the vessels, which had been invited on 15.8.34 were received on 25.9. and the vessels ordered on 13.12.34. The tender prices* were as follows.

* Plus Armament and Equipment supplied by the Admiralty.

NAME	BUILDER	HULL £	MACHINERY £	AUX MACH £	TOTAL PRICE £
HERO	Parsons Marine	106,235	134,033	9,590	249,858
HEREWARD	Steam Turbine Co. Ltd.*	105,935	134,066	9,590	249,591
HASTY	Denny	104,970	134,646	8,995	248,611
HAVOCK		104,680	134,795	8,995	248,470
HOSTILE	Scotts	107,750	136,717	8,915	253,382
HOTSPUR		107,405	136,717	8,915	253,037
HUNTER	Swan Hunter & Wigham Richardson (Wallsend Slipway)	106,870	136,778	9,519	253,167
HYPERION		106,600	135,347	9,519	251,466

* Hulls sub-contracted to Vickers-Armstrongs Ltd. (Tyne).

HEREWARD was fitted with an experimental twin 4.7" mounting forward in B position, with 'A' gun omitted and when inclined was 10 tons heavier than estimated. Welding was used during the construction of HOTSPUR.

BUILDING DATES

NAME	LAID DOWN	LAUNCHED	COMPLETED
HARDY	30.5.35	7.4.36	11.12.36
HERO	28.2.35	10.3.36	21.10.36
HEREWARD	28.2.35	10.3.36	9.12.36
HASTY	15.4.35	5.5.36	11.11.36
HAVOCK	15.5.35	7.7.36	16. 1.37
HOSTILE	27.2.35	24.1.36	10. 9.36
HOTSPUR	27.2.35	23.3.36	29.12.36
HUNTER	27.3.35	25.2.36	30. 9.36
HYPERION	27.3.35	8.4.36	3.12.36

HARDY is well known for her loss during the First Battle of Narvik on 9.4.40 and the subsequent award of a posthumous V.C. to her commander, Captain Warburton-Lee. (W.S.P.L.)

HARDY (H87)

HARDY, after working up at Portland, left for Mediterranean on 6.1.37 to join the 2nd D.F. as leader, where she superseded HOSTILE, which had been acting in this role. HARDY was to remain leader of this flotilla until her loss. During her period of service in the Mediterranean she spent between 5-6.37, 9-12.37, 3-8.38 and 2-3.39 either based at Gibraltar or off the Spanish Mediterranean coast on Non-Intervention duties. On 14.5.37 she stood by HUNTER after the latter had been mined off Almeria. 10 days later she was lying at Palma when the port was bombed by Republican aircraft but she was not damaged.

HARDY returned to the U.K. and refitted at Devonport between 2.6-29.7.39. Although war was imminent she returned to the Mediterranean and on 3.9.39 she was lying at Malta.

The Flotilla remained in the Mediterranean until 5.10.39 when HARDY led a division of her flotilla from Gibraltar to Freetown on anti-raider duties until 13.1.40. HARDY and her sisters were used as escorts for the major vessels that had been organised into hunting groups, searching for the ADMIRAL GRAF SPEE. When the latter was sunk, HARDY returned to the U.K. where she arrived at Plymouth on 25.1.40.

After docking and de-gaussing at Devonport she left on 12.2.40 for Greenock and convoy escort duties. She continued on these duties, based at Greenock, until 9.3.40 when she joined the Home Fleet at Scapa.

On 9.4.40, the 2nd D.F., whilst patrolling off the Lofoten Islands, received orders to proceed to Narvik to prevent German troops from landing there. Captain (D) of the Flotilla, Captain B.A.W. Warburton-Lee decided to attack the German squadrons in Ofot Fjord in spite of the Germans being in greater strength than expected. Initially the action went well, with HARDY torpedoing and blowing off the stern of the destroyer WILHELM HEIDKAMP. Another destroyer — the ANTON SCHMITT — was sunk and three others damaged, whilst six of the eight merchant ships present were also damaged. However, whilst returning down the Ofot Fjord, the Flotilla encountered five large German destroyers.

HARDY was torpedoed in the engine-room and run aground on the south shore of the Fiord in position 68°23'N 17°06'E. Survivors from HARDY were landed at Ballangen, Norway. HUNTER was also lost in the action, whilst HOTSPUR was damaged and escorted by HOSTILE and HAVOCK to Skjael Fjord. Casualties on HARDY were her Commanding Officer Captain Warburton-Lee, one other officer and 14 ratings killed with two other ratings missing. Warburton-Lee was posthumously awarded the Victoria Cross.

The wreck of HARDY was still visible as late as 1963.

A fine view of HASTY at speed. She had a fine record in the Mediterranean, but was finally lost on 13.6.42 during Operation 'Vigorous'. (Imperial War Museum A8654)

HASTY (H24)

On completion on 11.11.36, HASTY joined her sisters in the 2nd D.F. in the Mediterranean until 6.39, when she returned to the U.K. to refit at Devonport. In late 7.39, she re-commissioned and returned to the Mediterranean to work-up before spending 9.39-10.39 on escort duties between Port Said and Gibraltar. However, in 10.39 the First Division of the Flotilla — HARDY, HASTY, HERO and HOSTILE — were ordered to Freetown to join Force 'K' to undertake searches in the South Atlantic for German surface raiders and blockade runners for the remainder of the year. During 1.40, HASTY was ordered to the U.K. for docking and on 12.2.40 on her way home, she intercepted and captured the German blockade runner MOREA in the Western Approaches.

After completing her refit HASTY joined the Home Fleet and on 19-20.3.40, in company with the destroyer VIVIAN, she escorted into Newcastle the destroyer JERVIS which had been damaged in a collision with a Swedish merchant ship in the North Sea. Weather damage prevented her appearance at the First Battle of Narvik, but she withstood dive bombing whilst helping to evacuate the British Expeditionary Force from Andalsnes between 29.4-2.5.40.

A fortnight later, on 16.5.40, the reconstituted 2nd D.F. left Plymouth for the Mediterranean, where HASTY was to remain until her loss some two years later. During 1940, HASTY participated in a convoy operation to Malta, the Battle of Calabria and the action of 19.7.40 off Cape Spada, when the Italian cruiser BARTOLOMEO COLLEONI was sunk by the cruiser SYDNEY.

On 2.10.40 HASTY and HAVOCK sank the Italian submarine BERILLO off the Egyptian coast, capturing the vessel's C.O., 6 officers and 40 ratings. The next month HASTY was one of the escorts for the aircraft carrier ILLUSTRIOUS when the latter's aircraft attacked the Italian battlefleet at Taranto on 11.11.40, On 18.12.40 HASTY escorted the British squadron that bombarded the Albanian port of Valona.

1941 saw HASTY heavily engaged once more. On 16.1.41, she was one of the escorts for the aircraft carrier ILLUSTRIOUS when that ship was badly damaged by air attack off Malta. She participated in the battle of Matapan on 27.3.41 and a month later on 21.4.41 acted as a minesweeper with HOTSPUR, HERO and HAVOCK for the 3rd Battle Squadron which bombarded Tripoli. She then participated in the evacuation of Crete during 5.41 and a month later supported the army in the invasion of Syria and Lebanon. On 21-22.5.41 HASTY in company with the cruisers DIDO, ORION and the destroyers JANUS, KIMBERLEY and HEREWARD sank three steamers and 12 caiques off western Crete.

HASTY spent most of the last six months of the year undertaking no less than ten 'runs' between Alexandria and Tobruk. Between 4-12.41 the Royal Navy lost 25 vessels with 9 others damaged on this operation. On 23.12.41 HASTY stood by the landing ship Infantry (L.S.I.) GLENROY after the latter had been hit by an aerial torpedo during these operations. During 12.41 she escorted the transport BRECONSHIRE from Alexandria to Malta and then on 23.12.41 whilst escorting empty store ships from Tobruk, with

HOTSPUR sank U79 in postion 32°15'N 25°19'E, the first U-Boat sunk in the Eastern Mediterranean. Further escort duties to Malta followed in 1-2.42 before HASTY joined the 22nd D.F. on 24.2.42 on the re-organisation of the Destroyer Flotillas.

HASTY's epic career continued with her participation in the bombardment of Rhodes on 15.3.42, the second battle of Sirte on 20.3.42 where she received heavy weather damage and culminated in her final action protecting convoy MW11 from Alexandria to Malta on 13.6.42 as part of Operation 'VIGOROUS'. This convoy of 10 merchant ships escorted by 26 warships met continuous air, surface and submarine attack. On hearing that the enemy surface fleet would join the action, Admiral Harwood ordered the convoy to retire at 02.00 hours on 15.6.42. At 03.55 hours, German S-boats were sighted at a range of 1000 yards on the port bow. A boat was illuminated by NEWCASTLE, but it turned to port and torpedoed NEWCASTLE. HASTY gave chase and only returned to NEWCASTLE when it was thought that the S-Boat was of no further danger.

At 04.40 hours, when HASTY was on the starboard beam of NEWCASTLE attempting to get into the correct screening position, a torpedo fired by the S-boat (S55) was seen approaching the port quarter. The torpedo struck HASTY on the port side abreast 'A' gun. All the structure forward of 'A' gun was blown away, the forward boiler room was set on fire and both boiler rooms were flooding slowly. HASTY could steam astern but not ahead. HOTSPUR took off HASTY's ship's company and sank her by torpedo. The casualty roll was 12 killed and 1 died of wounds.

HAVOCK (H43)

HAVOCK was a member of the 2nd D.F. in the Mediterranean between 2.37-8.39 and operated off the Spanish Mediterranean coast between 5-9.37. On 1.9.37 whilst on these duties, she was unsuccessfully attacked by the Italian submarine IRIDE in position 38°46'N 00°31'E. She refitted at Gibraltar 19.10-13.11.37 and again repaired at Gibraltar 16.4-6.5.38 after hitting a quay wall.

HAVOCK, pictured off Chatham, probably after her short refit at Sheerness during 8.39.
(National Maritime Museum N23170)

She returned to the U.K. during 8.39 and received a limited refit at Sheerness 15-26.8.39 and on that day she left for Gibraltar. At Gibraltar she received orders to join the South Atlantic Station and left on 30.8.39 for Freetown, where she arrived on 4.9.39 and joined other members of the 2nd Flotilla there. She remained on station until 11.39, when she returned to the U.K. to refit at Sheerness 18.12.39-23.3.40.

HAVOCK then rejoined the 2nd D.F. now with the Home Fleet at Scapa on 27.3.40. She participated in the First Battle of Narvik when she forced ashore the German ammunition ship RAUENFELS, which later blew up. At the end of the action HAVOCK and HOSTILE escorted the disabled HOTSPUR to Skjael Fjord. HAVOCK reported the capture of 16 prisoners and also that one of her guns was out of action.

After completing repairs, HAVOCK and HYPERION joined the Nore Command and on 10.5.40 they engaged German forces at Waalhaven Aerodrome in Holland. The two sisters then patrolled off the Hook of Holland and survived heavy air-attacks. HAVOCK later picked up survivors from the ferry PRINSES JULIANA and returned them to the Hook, where she picked up naval and military demolition parties.

On 16.5.40 she sailed as one of the reinforcements for the Mediterranean Fleet and arrived at Malta six days later, rejoining the 2nd D.F., with which she was to serve until 2.42.

On 19.7.40 off Cape Spada, Crete, whilst on a sweep with her sister HYPERION and the Australian cruiser SYDNEY she sighted the Italian cruiser BARTOLOMEO COLLEONI and in the action that followed the Italian ship was sunk. However, whilst proceeding south after this action, HAVOCK was bombed and her boiler room was flooded. Repairs were completed at Suez between 29.7-15.9.40.

On 2.10.40 in company with HASTY she surprised the surfaced Italian submarine BERILLO which after being gunned and depth charged surrendered and scuttled herself. After this action, HAVOCK continued on normal destroyer duties until the eruption of the Greek and Cretan campaigns. However between 22.12.40-20.2.41 she was under refit and repair at Malta.

On 23.5.41 she arrived at Alexandria, after being damaged in an air attack by dive-bombers following a night patrol off Heraklion, Crete. Repairs at Alexandria were completed on 16.6.41. She then operated off the Lebanese coast during the same month, before undertaking eight runs to Tobruk between 7-10.41. On 21.10.41 she arrived at Alexandria with damage to both her shafts and propellers. She was again repaired at Alexandria 21.10-4.12.41. Two weeks later she was damaged by splinters whilst lying at Malta and was out of action for a further three days. HAVOCK participated in Convoy Operation MF3, but was detached to escort the damaged transport THERMOPYLAE from Benghazi to Alexandria. The HAVOCK, in difficult conditions, rescued some of THERMOPYLAE's crew when the latter was sunk by bombing.

HAVOCK joined the 22nd D.F. during 2.42 and a month later escorted a further convoy from Alexandria to Malta, which resulted in the 2nd Battle of Sirte on 22.3.42 in which she received extensive splinter damage. On assessment at Malta the next day, it was found that HAVOCK required repairs to her boiler-room and hull members had to be strengthened. No 3 boiler room was flooded and probably damaged as well. It was estimated that damage repairs would take 14 days.

However, HAVOCK and the cruiser PENELOPE had become the primary target for enemy air-raids whilst in dock and on 3.4.42 some plating and framing was damaged by near misses. It was then decided that HAVOCK would leave Malta as soon as possible.

On the night of 5.4.42 she left Malta independently for Gibraltar, but the next day she ran aground off Keleba Lighthouse on the Tunisian coast and was destroyed by her crew. One rating was killed. Her crew and the passengers on board were interned by the French, but following the invasion of North Africa, they were all released. She lies in position 36° 52' 27" N 11° 08' 30" E.

HAVOCK received damage at the 2nd Battle of Sirte and repairs were started at Malta, but were stopped by air attacks and HAVOCK then made a dash for Gibraltar, but ran aground off the Keleba Lighthouse on 5.4.42.
(Imperial War Museum HU2870)

HEREWARD (H93)

On commissioning, HEREWARD undertook trials of the newly designed twin 4.7" mounting that was due to be fitted to the new 'Tribal' class destroyers then under construction. The mount was fitted in 'B' position and was tested at Gibraltar between 1-3.37. She had previously undertaken torpedo discharge trials at Portland between 11-18.1.37 before leaving for Gibraltar.

On completion of the gunnery trials, HEREWARD exchanged the twin mount for two single mounts at Portsmouth between 16.3-18.5.37 and two days later left to join the 2nd D.F. of the Mediterranean Fleet. She immediately began operations off the Spanish Mediterranean coast until 9.37. HEREWARD then refitted at Malta 30.9-30.10.37 and operated with her flotilla in Mediterranean waters, before returning to the U.K. to refit at Portsmouth 7.6-26.7.39. She was, however, on station in the Mediterranean on 3.9.39 and remained in the Mediterranean until 10.39, when she transferred to the South Atlantic Command and was based at Freetown 15.10-11.11.39 and then at Trinidad 20.11.39-23.1.40. The highlight of her Atlantic service was to escort the aircraft carrier ARK ROYAL, as part of a hunting group. She boarded the German merchant ship UHENFELS some 300 miles west-south-west of Freetown and took her into Freetown as a prize. HEREWARD also blockaded the German merchant ship ARAUCA to prevent her from leaving Port Everglades, Florida.

HEREWARD pictured in early 1937, when undertaking trials of the new twin 4.7" mounting destined for the "Tribals" then building. The gun mounting was subsequently removed on the successful completion of these trials.

On 23.1.40 HEREWARD and HUNTER left Bermuda for Halifax as escort for the battleship VALIANT. However, on arrival at Halifax on 27.1.40, HEREWARD was docked for weather damage repairs until 16.2.40. She left Halifax twelve days later as escort for the cruiser ORION conveying the Governor-General's ashes to the U.K. Further defects and the need to be de-gaussed meant that she remained at Portsmouth from 11.3-12.4.40. She, therefore, missed the First Battle of Narvik.

After working up, HEREWARD joined the Home Fleet on 8.5.40, but was one of eight destroyers detached to Harwich to meet the threat that had then arisen when the Germans invaded the Low Countries on 10.5.40. The next day, she was one of a group of destroyers that escorted trawlers into Scheveningen to evacuate British nationals. Two days later, she embarked the Queen of the Netherlands at the Hook of Holland and transported her to Harwich before returning to other evacuation duties.

She was then allocated to the Mediterranean Fleet and was one of a group of destroyers that left Plymouth on 17.5.40 for the Mediterranean. She re-joined the 2nd D.F. at Alexandria on 24.5.40. She was to be heavily engaged in the next months — she participated in the actions off Calabria during 7.40, the Cape Spartivento action on 27.11.40, and undertook the escort of convoys as part of Operation "COLLAR". Between 10-12.12.40, she harassed road transport during the Italian retreat after the battle of Sidi Barrani.

The next day, HEREWARD and HYPERION sank the Italian submarine NAIADE, some 30 miles north of Sidi Barrani, rescuing her entire crew. On 19.12.40, she participated in the bombardment of Valona, and on 25.12.40 she went to the aid of a convoy that had been attacked by the German cruiser ADMIRAL HIPPER. She returned to Gibraltar on 29.12.40 with three ships from the convoy.

On 6.1.41, HEREWARD and three other destroyers escorted the cruiser BONAVENTURE when she left Gibraltar as part of Operation 'EXCESS' and four days later, when 12 miles south-east of Pantellaria two Italian warships were engaged and HEREWARD sank one of them, the VEGA, with a torpedo. On the successful conclusion of the operation on 13.1.41, HEREWARD escorted the armed boarding vessel CHAKLA into Benghazi. Nine days later, with DECOY, she carried a commando force to the island of Castellorizo, off the Turkish coast. The operation was a disaster, as most of the occupying force was later captured.

HEREWARD next participated in the Battle of Matapan on 28.3.41 and then escorted convoys to Greece and after 24.4.41 evacuated troops to Alexandria and Crete. HEREWARD was in the thick of the fighting around Crete and on 21.5.40, she sank a number of caiques conveying German troops, some 25 mies off Suda Bay. Seven days later, at 06.00 hours on 28.5.41, the cruisers ORION, AJAX and DIDO with six destroyers including HEREWARD sailed for Heraklion to evacuate the garrison. The evacuation was successfully completed and at 03.20 hours the next day, the vessels set sail. The destroyer IMPERIAL was then sunk by bombing and at 06.50 hours HEREWARD was hit and her speed reduced. To avoid imperilling the whole force, it was decided to leave HEREWARD behind. She was sunk by further air attacks close to the Cretan coast. Four officers and 72 ratings were lost and 6 officers and 83 ratings were captured. The vessel lies in position 35° 20' N 26° 20' E.

HERO/H.M.C.S. CHAUDIERE (H99)

On completion on 21.10.36, HERO was allocated to the 2nd D.F. of the Mediterranean Fleet and operated on normal destroyer duties and detachments off the Spanish coast until her return to the UK to refit during 6.39.

HERO at Malta on 5.1.37, whilst serving with the 2nd D.F. of the Mediterranean Fleet.

On 26.7.39, she recommissioned for service with the Mediterranean Fleet and on 3.9.39 was lying at Malta. She remained in the Mediterranean for the next month until the intentions of Italy were known and when Italy's neutrality was secure, HERO left Gibraltar on 5.10.39 for service in the South Atlantic based at Freetown. She operated as a unit of raider hunting groups until 1.40 when she returned to Gibraltar. During 12.39 she had been based at Pernambuco. She had a brief stop for repairs at Gibraltar, and then refitted at Portsmouth between 15.2-16.3.40 before rejoining the Home Fleet's 2nd D.F. for service in Norwegian waters.

HERO was quickly in action; between 5-8.4.40 she participated in Operation 'WILFRED' a minelaying operation in Vest Fjord and two days later she sank U.50 with depth charges off Muckle Flugga. On 13.4.40, HERO was escort for the battleship WARSPITE with the destroyers ICARUS, FOXHOUND, KIMBERLEY, FORESTER, BEDOUIN, PUNJABI, ESKIMO and COSSACK during the 2nd Battle of Narvik, when eight German destroyers were sunk avenging the loss of her sisters HARDY and HUNTER, three days earlier. She then remained in Norwegian waters for the next month.

On 17.5.40 she was one of the reinforcements sent from the UK to re-consitute the 2nd D.F. in the Mediterranean. Her next six months service was to be action packed.

On 17.7.40 she participated in the action off Cape Spada, when the Australian cruiser SYDNEY sank the Italian cruiser BARTOLOMEO COLLEONI. On 23.8.40, whilst on passage across the Mediterranean HERO and MOHAWK picked up the survivors from her sister ship HOSTILE, which had been mined off Cape Bon, and returned her survivors to Malta. She later joined convoy operation 'HATS' that successfully passed through the Mediterranean under air attack.

HERO refitted at Malta during 11.40 before being a participant in the action off Cape Spartivento on 27.11.40. HERO then joined Force 'H' and the 13th D.F. at Gibraltar and when the troop convoy WS5A was scattered after being attacked by the German cruiser ADMIRAL HIPPER on 25.12.40, she was one of the vessels sent out to round up and protect the scattered ships.

On New Year's Day 1941, HERO and other vessels of the 13th D.F. intercepted a Vichy French convoy of four vessels off Oran and escorted them into Gibraltar. Five days later HERO, with JAGUAR, HEREWARD and HASTY escorted part of the 'EXCESS' convoy and rejoined the Mediterranean Fleet. On 20.1.41, HERO was one of the escorts of the cruisers ORION and BONAVENTURE during an abortive bombardment of Tobruk. She survived the Greek and Cretan debacles without damage. 5.41 saw HERO's participation in Operation 'TIGER', a supply operation to Malta and on the night of 22/23.5.41 HERO and DECOY evacuated the Greek King and his court from Crete to Alexandria. The next day the two ships carried 600 special service troops to Suda Bay and returned safely.

6.41 saw HERO operating off the Syrian Coast and between 10-13.7.41 she undertook her first stores run to Tobruk. She was to be on this dangerous duty for the next three months, interrupted by her participation in a diversionary operation with the battleship QUEEN ELIZABETH during the passage of the convoy WS11 during Operation 'HALBERD' in 9.41.

On 25.10.41 when proceeding to Tobruk with the minelayer LATONA and the destroyers ENCOUNTER and EXPRESS the ships were subjected to a succession of air attacks from 14.32 hours onwards. At 20.10 hours LATONA was hit by a bomb, disabled and set on fire. 40 minutes later HERO was alongside LATONA's port bow and secured bow to bow. Almost simultaneously with HERO securing alongside, she was near-missed by three bombs about 10ft on her starboard side, one abreast No.3 boiler, the second abreast of the 3" gun at station 128 and the third just cleared the stern. The ship whipped violently three times, but as initial reports indicated that no major damage had been caused, the evacuation of LATONA's crew continued.

The plating on HERO's starboard side was indented from station 87 to station 122 and from 4ft to 13ft below the upper deck. Rivets were sprung and her fighting efficiency impaired as her maximum speed was reduced to 29 knots and her manoeuvrability affected. She was repaired at Alexandria and then operated on escort duties in the Eastern Mediterranean.

During 2.42 she formed part of the 22nd D.F. with SIKH, ZULU, LEGION, LANCE, LIVELY, HAVOCK and HASTY and by 9.42 was to be the sole survivor of this Flotilla, such were the losses of Royal Navy destroyers at this time. HERO participated in the Second Battle of Sirte on 22.3.42 as escort for MW10. On 28.5.42, whilst on patrol with ERIDGE and HURWORTH, she sank U568 after a fifteen hour hunt north-east of Tobruk and rescued 42 survivors.

On 30.6.42 the depot ship MEDWAY, on passage from Alexandria to Haifa, was torpedoed by U732 and sank in fifteen minutes. HERO and ZULU picked up 1105 survivors between them. HERO was in dire need of a refit, but continued on station, escorting convoys between Cyprus and Alexandria and other Levant ports until the end of the year. The highlights of this part of her career were the rescuing of some 1100 survivors from the torpedoed troop transport PRINCESS MARGUERITE on passage between Alexandria and Cyprus on 17.8.42 and some 2 months later on 30.10.42, with four other destroyers and a Wellesley aircraft of 42 squadron, she hunted down and sank U559 some 60 miles north-east of Port Said.

By 12.42 a major refit could be postponed no longer and she returned to the UK via the Cape. She then refitted at Portsmouth 4-11.43 and was converted to an escort destroyer. On 15.11.43 she was transferred to the Royal Canadian Navy as a gift and re-commissioned the same day as CHAUDIERE.

After working-up at Scapa from 12.43, CHAUDIERE joined Escort Group C2 at Londonderry during 2.44. She was immediately in action protecting convoy HX228 some 480 miles west of Cape Clear, Ireland. After a 32 hour hunt, U744 was forced to the surface and surrendered on 6.3.44. She, however, proved impracticable to tow and the submarine was torpedoed and sunk by ICARUS.

On 25.4.44 CHAUDIERE joined the 11th Escort Group and protected invasion traffic from the predations of those U-boats based on the Biscay ports. CHAUDIERE participated in the sinking of no less than 3 U-boats-U678 with OTTAWA (II) and the frigate STATICE on 6.7.44 off Beachy Head and U621 and U984 again with OTTAWA (II) and KOOTENAY on 18.8.44 and 20.8.44 in the Channel. During 10.44, the escort group transferred operations to Icelandic waters for the next month. However on 19.11.44, CHAUDIERE left Londonderry as an extra escort for convoy ON267 and arrived at Sydney, Cape Breton ten days later, where she was to refit.

This was to be CHAUDIERE's last voyage, as she did not enter refit at Sydney until 22.1.45 and was still refitting on V.E. day. On inspection, she was found to be in the poorest condition of any of the surviving destroyers in Canadian service. She was declared to be surplus to requirements on 13.6.45 and finally paid off at Sydney on 17.8.45. She was later sold for demolition to the Dominon Steel Company, but was not broken up until 1950 at Sydney. Thus passed a vessel that had survived the Mediterranean and participated in the destruction of no fewer than six U-boats!

HOSTILE was to have a short career on her return to the Mediterranean in 1940. She was mined off Cape Bon on 23.8.40.

HOSTILE (H55)

Completed on 10.9.36, HOSTILE relieved CRESCENT in the 2nd D.F. in the Mediterranean the following month. She became the temporary leader of the flotilla until HARDY's arrival during 1.37. Between 3-6.37 and 14.8-22.9.37 she patrolled off the Spanish Mediterranean coast. On 23.8.37 she aided the s.s. NOEMIJULIA and escorted her into Port Vendres. HOSTILE refitted at Gibraltar between 17.11-15.12.37.

1938 saw HOSTILE undertake patrols off the Spanish Coast between March and August. Between 1-5.39, HOSTILE operated in the Western Mediterranean and was stationed off Barcelona between 26.1-8.2.39 during the final days of the Spanish Civil War.

HOSTILE arrived at Sheerness on 31.5.39 to refit, work which was completed on 26.7.39. She immediately returned to the Mediterranean and was lying at Malta when war broke out. After remaining in the Mediterranean for several weeks, she was ordered to the South Atlantic and was based at Freetown between 15.10.39-13.1.40 on anti-raider patrols with Force 'K'.

On her return to the U.K. she repaired defects at Chatham between 26.1-29.3.40 before escorting the battle-cruisers REPULSE and RENOWN on passage from Plymouth to Greenock between 1-3.4.40. HOSTILE then rejoined the 2nd D.F. for service with the Home Fleet. Between 31.3-8.4.40 with FEARLESS she escorted the cruiser BIRMINGHAM searching for German fishing vessels along the Norwegian coast.

HOSTILE escaped relatively unscathed from the first battle of Narvik on 10.4.40 and helped to tow her disabled sister HOTSPUR away for repairs.

After further service off Norway, HOSTILE repaired defects at Rosyth between 27.4-4.5.40 and twelve days later she left Plymouth for the Mediterranean as one of the group of reinforcements from the U.K. She was to serve with the re-constituted 2nd D.F. for scarcely three months but participated in the action off Calabria on 9.7.40.

On 22.8.40 the destroyers HERO, HOSTILE, MOHAWK and NUBIAN were ordered to join Force 'H' at Gibraltar. At 02.17 hours on 23.8.40 when 176 miles and in a position 125 degrees from Cap Bon Lighthouse, an explosion occurred under HOSTILE abaft the engine room. The ship had struck a mine in field ANS which had been laid by Italian destroyers a fortnight before.

The explosion occurred in the region of frame 130 on the port side beneath the ship and made a hole in the side plating, a large part of which folded outwards. The upper deck bulged upwards in the vicinity of the after torpedo tubes which were blown overboard as were two of the crew. The Engineering Officer stated that the ship's back was broken and the two ends of the vessel were held together by the upperdeck plating. The upperdeck was awash on the starboard side very soon after the explosion. When the torpedo to scuttle the ship was fired HOSTILE's upperdeck was awash as far forward as the searchlight platform.

Five of her complement were killed: an officer sleeping in the Captain's cabin, one rating, two torpedo men and a stoker. Three others were injured.

MOHAWK closed HOSTILE and took off survivors and HOSTILE was abandoned at 03.35 hours. HERO then closed HOSTILE and fired a torpedo which struck abreast of the forward boiler room. There was a large explosion and within 40 seconds HOSTILE had sunk so that only 40' of her bows remained above water. Subsequently HERO fired a second torpedo to sink her bow and when at 04.15 hours HERO proceeded to rejoin MOHAWK nothing remained of HOSTILE. She lies in position 36° 54' 48" N 11° 20' 45" E.

HOTSPUR (HO1)

On commissioning, HOTSPUR joined her sisters in the 2nd D.F. in the Mediteranean. HOTSPUR operated off the Spanish Mediterranean coast 8-9.37, 7-8.38 and at Gibraltar 2-3.39. She refitted there between 16.12.37-17.1.38 and arrived at Sheerness to refit during 8.39.

This refit was terminated due to the threating international situation and on 26.8.39, HOTSPUR sailed for the Mediterranean. However on her arrival at Gibraltar she was diverted to Freetown and operated with other vessels of the 2nd D.F. on anti-raider patrols in the South Atlantic until 22.10.39. She then operated in the West Indies for the next two months.

She returned to the U.K. on 18.1.40 and refitted at Sheerness until 6.3.40. A week later she was operating with the 2nd D.F. with the Home Fleet at Scapa. At the Narvik action of 10.4.40, she received seven direct hits which put her depth charge throwers and her rangefinder out of action. All electrical circuits were put out of action as was her No 2 boiler and her seaworthiness was impaired by shell holes. She then collided with her sister HUNTER and the two vessels when locked together drew further enemy fire and 18 of her crew were killed. After separation and temporary repairs at Skjael Fjord HOTSPUR steamed back to Chatham by the west coast and repairs were effected between 2.5.-16.7.40.

HOTSPUR sailed for Gibraltar on 24.7.40 and joined the 13th D.F. of the North Atlantic Command. On 18.10.40 HOTSPUR with GALLANT and GRIFFIN sank the Italian submarine LAFOLE in position 35° 50'N 02° 53'W north of Melilla. HOTSPUR, whilst ramming the submarine, received considerable damage to her hull for a distance of about 45ft from her stem. Temporary repairs were completed at Gibraltar between 22.10-20.11.40. She then transferred to Malta where permanent repairs were undertaken between 29.11.40-20.2.41. She then joined the Mediterranean Fleet's 2nd D.F. for an action packed year.

After participating in the Cape Matapan action, she undertook convoy escort duties to Greece and Crete in 4-5.41 and, on 7.5.41, with HAVOCK and IMPERIAL, she escorted the cruiser AJAX on her bombardment of Benghazi. During the Crete debacle, she was ordered to sink the IMPERIAL, after the latter had been damaged by a near-miss on 29.5.41.

HOTSPUR was then engaged on the highly dangerous Tobruk convoy run from Alexandria between 7-11.41. On 29.11.41, she was attacked by two torpedo bombers in position 31° 10' N 28° 45' E, but was undamaged. HOTSPUR and GRIFFIN then escorted the cruiser NAIAD when she bombarded Derna. Later, on 23.12.41 HOTSPUR, with HASTY, sank U79 in position 32° 15'N 25° 19'E.

On 23.2.42, on the re-organisation of the Mediterranean Flotillas HOTSPUR was transferred to the Eastern Fleet but nominally remained a member of the 2nd D.F. However, she returned to the Mediterranean to

HOTSPUR seen late in the war, as an A/S escort and showing many war modifications.

escort the unsuccessful 'VIGOROUS' convoy to Malta between 13-21.6.42. During this operation she had the melancholy task of sinking her sister HASTY, after the latter had been damaged by an S-boat torpedo on 15.6.42.

HOTSPUR then returned to the Indian Ocean where she operated until 1.43. She arrived at Freetown from Simonstown on 14.2.43 and was briefly attached to the 4th D.F. She arrived at Plymouth on 27.2.43 before undergoing a refit at Sheerness between 1.3-31.5.43. At this time she was also converted to an escort destroyer. After working-up at Tobermory during 6.43 she joined the Western Approaches Command on convoy escort duties until 10.44. She operated with Escort Group C4 until 6.44 and then the 14th Escort Group.

She then refitted at Barrow between 31.10.44-9.3.45 and this was followed by local escort duties in the Irish Sea until V.E. Day. After a brief period with Rosyth Escort Force, HOTSPUR operated with the Londonderry Training Squadron between 8.45 and 6.46. She then joined the 4th Escort Group until 2.47. She refitted at Portsmouth during 2-3.47 before joining the 3rd Escort Flotilla at Portland 6.47-2.48.

HOTSPUR on 12.5.47 after a refit and prior to joining the 3rd Escort Flotilla at Portland. A year later she became the Dominican Republic's TRUJILLO.

HOTSPUR was approved for scrapping during 11.47 and ordered to enter Category 'C' Reserve on 20.1.48. She left the reserve during 6.48, but remained laid-up at Portsmouth until 23.11.48 when she was sold to the Dominican Republic and re-named TRUJILLO (see FAME for details of sale).

She was to serve in the Dominican Navy for some 25 years, being named DUARTE in 1962 following the deposing of the Trujillo family and was not scrapped until the 1970's.

HUNTER (H35)

HUNTER's career nearly came to an abrupt end only eight months after commissioning when, on 13.5.37 she was mined off Almeria, on the south coast of Spain. The minefield had been laid by the Nationalist M.T.Bs S2 and S4. Immediately after the mining, HUNTER was towed away from the danger area by the Republican destroyer LAZAGA. She received damage on her port side and was later towed, in near sinking condition by the cruiser ARETHUSA with HARDY standing by. She arrived in at Gibraltar on 15.5.37, escorted by IMOGEN and ICARUS.

HUNTER well down forward after being severely damaged by a mine off Almeria on the Spanish Mediterranean coast on 13.5.37. Subsequent repairs at Malta were not completed until 11.38. (Imperial War Museum HU49098)

Temporary repairs were undertaken at Gibraltar between 15.5-18.8.37. She paid off into dockyard hands on 29.5.37 and was replaced by ACTIVE. She was towed to Malta 19-21.8.37 for permanent repairs, which were not completed until 10.11.38.

HUNTER circa 10.38 following her re-building after being mined. She is running trials without her fore funnel. (W.S.P.L.)

On 23.1.39, HUNTER left Malta for service with the 2nd D.F. and remained in the western Mediterranean until 6.39. Between 24.6-4.7.39 repairs to oil leaks were undertaken at Malta. She arrived at Plymouth on 13.8.39 to give leave and refit between 15-27.8.39.

On 3.9.39 HUNTER was on passage to Freetown to join one of the raider hunting groups then being formed. She was based at Freetown from 4.9.-23.10.39 and then operated in the West Indies between 4.11-4.12.39. She arrived at Halifax on that day, but was to be based at Bermuda until 23.1.40.

HUNTER and HEREWARD served as escorts for the battleship VALIANT which had been working up there and with the cruiser ENTERPRISE left Bermuda on 23.1.40 for Halifax, where they arrived on 30.1.40. HUNTER then made passage to Plymouth where she arrived on 8.2.40 and immediately refitted at Falmouth until 9.3.40.

After escorting the battleship WARSPITE from Greenock, she arrived at Scapa on 17.3.40 for service with the Home Fleet's 2nd D.F. HUNTER with the rest of the flotilla left Scapa on 10.4.40 for Norway where she participated in the First Battle of Narvik. HUNTER was damaged by the gunfire of the German destroyers and was lost following a collision with her sister HOTSPUR. Eight officers and 99 ratings were killed and five ratings died later of wounds. Two officers and 44 ratings were taken aboard the German destroyer ERICH GIESE and were landed at Narvik, where they were held as prisoners until 13.4.40, when they were released by the Germans and sent to Sweden. The wreck of HUNTER lies in 68° 20' N 17° 04' E.

HYPERION immediately after launch on 8.4.36. (Swan Hunter Ltd.)

HYPERION (H97)

On commissioning, HYPERION joined her flotilla mates in the Mediterranean Fleet's 2nd D.F. during 1.37 and remained on-station with her Flotilla until 8.39. She spent between 3-6.37, 8-9.37, 3-9.38, 10-12.38 and 2-4.39 off the Spanish Mediterranean Coast on Non-Intervention patrol duties. She refitted at Malta between 30.9-30.10.37 and arrived at Portsmouth to refit during 8.39. However, some minor works were completed at the Dockyard between 16-27.8.39, before she sailed for the South Atlantic to be based at Freetown.

She was to remain at Freetown for the next two months on anti-raider duties, before transferring to Bermuda for similar duties until 1.40. She also performed blockade duties against German vessels lying in various Mexican and American ports.

On 19.12.39, HYPERION intercepted the German liner COLUMBUS (32,581 tons) off Cape Hatteras, which then scuttled herself. The German vessel's passage from Mexico had been reported in plain language by a number of U.S. Navy vessels.

HYPERION then returned to the UK and arrived at Portsmouth on 15.1.40. She refitted there between 25.1-6.3.40 before joining the Home Fleet's 2nd D.F. for service in Norwegian waters. With the cruiser CALCUTTA she escorted the destroyer ESCORT, which was towing the destroyer ECLIPSE to Sullom Voe from off the Norwegian coast, where the latter had been damaged by bombing.

Between 8-12.5.40 HYPERION operated off the Hook of Holland and the evacuation of British personnel. She was then ordered to the Mediterranean as a reinforcement, leaving Plymouth on 16.5.40 and arrived at Malta five days later. The final seven months of her existence with the 2nd D.F. were hectic. She participated in the action off Calabria on 9.7.40 and ten days later was one of the destroyers escorting the Australian cruiser SYDNEY when she sank the Italian cruiser BARTOLMEO COLLEONI off Cape Spada. HYPERION rescued some of the 525 survivors from the Italian ship. Later, she participated in the successful reinforcement convoy between 4-13.11.40. A month later HYPERION and HEREWARD sank the Italian submarine NAIADE off Bardia on 14.12.40. Eight days later, having just completed the escort of the battleship MALAYA to join Force 'H', HYPERION was mined between Cape Bon and Pantellaria. The field, codenamed 4AN, had been laid by the Italian destroyers UGOLINO VIVALDI, ANTONIO DA NOLI and LUCA TARIGO on the night of 7/8.10.40.

ILEX attempted to tow HYPERION but after the tow parted twice, with HYPERION settling aft, Captain Mack (Captain D2) decided that it would be impossible to clear Pantellaria before dawn and JANUS was order to sink her. ILEX took off the ship's company, but two of HYPERION's ratings were found to be missing and were presumed killed. She lies in 37°04' N, 11°31' E.

THE 1935 PROGRAMME
THE LEADER: INGLEFIELD

The Abyssian crisis was at its height when Cammell Laird, who were building the latest leader HARDY were requested on 18.10.35 to tender for a repeat of HARDY. The tender was returned on 26.10.35 and accepted 4 days later. The vessel was not actually ordered until 2.1.36, when the letter of 30.10.35 was confirmed. The tender price was £283,229 of which the hull was to cost £123,080, main machinery £151,867 and auxiliary machinery £8,282. This was £4,747 more than HARDY's tender price and indicates the higher profit levels of the shipbuilders. HARDY's final price was £285,989 with a profit of £22,995. INGLEFIELD's final price was £291,531 with a profit of £48,776.

When inclined on 20.7.37, INGLEFIELD's stability was found to be satisfactory with her water ballast tanks empty, so long as her displacement did not fall below 1,840 tons. Below this figure, all water ballast tanks had to be flooded. The design had finally reached the limit of its development potential.

THE DESTROYERS: THE INTREPIDS

It had originally been intended to order minelaying destroyers as part of the 1934 Programme, but orders had been delayed pending the results of the minelaying trials being conducted in the 2 E class destroyers. The following changes were incorporated in the design to allow the fitting of minelaying equipment:
(i) The fitting of sponsons aft.
(ii) The steering compartment was so arranged that the mine conveyor gear could be subsequently fitted.
(iii) The after superstructure was to be narrower than in ESK.
(iv) Torpedo loading derrick to have an increased overhang in order to embark mines.
(v) Paravane winch seats to be portable.
(vi) Alternative positions for depth charge throwers.
(vii) Whaler to be stationed on the forecastle, when employed as a minelayer.
(viii) Space on the bridge was to be set aside for mining instruments.

On 17.10.35 it was decided to order a repeat 'H' class destroyer class and the next day the following amendments to the design were made:
(i) Quintuple torpedo tubes to be fitted with a weight penalty of 10 tons.
(ii) The machine guns to be moved to the position of the 2 pounder pom-pom as in the "Tribals".

Delivery was requested by 30.4.37, but was not to be met for any of the vessels in the class because of lengthening order books and constraints in the manufacture of gun mountings. Hawthorn Leslie, J.S. White, John Brown and Yarrow were asked to return tender documents by 26.10.35. Letters of acceptance were despatched four day later—the days of leisurely tendering had gone! The tender prices were as follows:

TENDER PRICES (£)

VESSEL	BUILDERS	HULL	MACHINERY	AUX.MACHINERY	TOTAL COST
IMPULSIVE	J.S. White	109,190	141,180	8,955	259,325
INTREPID		108,830	141,180	8,955	258,965
IMOGEN	Hawthorn	107,970	139,931	9,016	256,917
IMPERIAL	Leslie	108,170	139,931	9,016	257,117
ICARUS	John Brown	107,600	138,857	9,015	255,472
ILEX		107,200	138,857	9,015	255,072
ISIS	Yarrow	109,200	141,264	8,907	259,371
IVANHOE		109,200	141,264	8,907	259,371

These vessels were between £6,000 and £11,000 more expensive than the H's.

WEIGHT PROBLEMS

By this time (10.35), the weight and stability problems of the 'H' class were becoming apparent and the D.N.C. informed the Controller that the removal of quintuple torpedo tubes and their replacement with quadruple mountings, was not acceptable as it would impair fighting efficiency and that the following measures to improve stability of the 'I' class were required instead:
(i) The flat to the magazine and shell rooms was to be made watertight and fitted with flooding arrangements. These compartments were to be flooded when the displacement reached the critical level.
(ii) 7lb and 5lb($\frac{1}{8}$ inch) plating was to be fitted in place of bullet proof plating on the bridge—a saving of 1.5 tons.
(iii) The vessels were to ship 30 tons of ballast to make the I's GM similar to that of the G's minimum figures. INGLEFIELD to be so fitted after HARDY had been inclined.

After ICARUS's inclining test, it was found that the vessels' stability ballasted was not worse than the 'H' class in light condition.

BUILDING PROGRAMME

NAME	LAID DOWN	LAUNCHED	COMPLETED
INGLEFIELD	29.4.36	15.10.36	25.6.37
IMPULSIVE	9.3.36	1. 3.37	29.1.38
ICARUS	9.3.36	26.11.36	1.5.37
ILEX	16.3.36	28. 1.37	7.7.37
IMOGEN	18.1.36	30.10.36	2.6.37
IMPERIAL	29.1.36	11.12.36	30.6.37
INTREPID	6.1.36	17.12.36	29.7.37
ISIS	5.2.36	12.11.36	2.6.37
IVANHOE	12.2.36	11. 2.37	24.8.37

INGLEFIELD on completion in 1937. She served with distinction before being lost to glider bomb attack off Anzio on 25.2.44.

INGLEFIELD (D02)

After completing, INGLEFIELD became leader of the 3rd D.F. serving in the Eastern Mediterranean during 1938-39. She was at Malta at the outbreak of war and with her Flotilla immediately sailed for Gibraltar on 5.9.39 for service in the Western Approaches Command.

Her first duty was to escort the aircraft carrier COURAGEOUS on an anti-submarine patrol which left Plymouth on 16.9.39. When COURAGEOUS was sunk she had, however, been detached to rescue survivors of the merchant ship KAFIRISTAN, sunk by gunfire by U53 in position 50°16'N 16°55'W on 17.9.39. INGLEFIELD returned at 01.00 hrs the next morning and organised an unsuccessful hunt for the submarine.

She was to remain leader of the 3rd D.F. until 6.42, and was employed with the remainder of her Flotilla as a hunting force for U-boats in the South West Approaches during 10.39. On 14.10.39 INGLEFIELD with IVANHOE, INTREPID and ICARUS sank U45 off Southern Ireland.

Five days later she started a refit at Liverpool which was completed on 11.11.39 when she returned to her duties with the 3rd Flotilla now a part of the Home Fleet. On 25.2.40 whilst escorting convoy HN14, INGLEFIELD sank U63 by gunfire after the submarine had been damaged by ESCORT, IMOGEN and the submarine NARWHAL in position 58°40'N 00°10'W. U63's entire crew of 4 officers and 21 ratings was captured.

INGLEFIELD was deeply involved in the Norwegian campaign and escorted the minelayer TEVIOTBANK which laid a minefield off Stadtland, Norway on 8.4.40 and participated in the evacuation of Andalsnes on the night of 1/2.5.40. She had a brief refit at Devonport between 19.5-3.6.40.

On 20.9.40 she intercepted the Vichy French cruiser GLOIRE and after the intervention of a Free French officer, the GLOIRE returned to Casablanca. Three days later she was involved in the unsuccessful Dakar operation during which she engaged a Vichy French submarine and was hit by shellfire from a shore battery. An old French 7.5" Lyddite high explosive shell fired at a range of about 6000 yards hit INGLEFIELD on the port side at frame 156 some 5ft below the upper deck and burst inside the ship. The cabin in which the shell burst was completely wrecked and one officer and six ratings were injured. The next day, however, INGLEFIELD and two other destroyers left the super destroyer L'AUDACIEUX burning after an engagement with them at Rufisque.

INGLEFIELD remained at Freetown for a month before returning to the UK as escort for convoy HG46 and on reaching London underwent a refit which was not completed until 20.1.41. She then returned to her duties as leader of the 3rd Flotilla and during the next nine months, as well as undertaking normal screening and escort duties, she participated in the raid on the Lofoten Islands 2-6.3.41, the unsuccessful search for the GNEISENAU and SCHARNHORST 19-23.3.41 and the BISMARCK action in 5.41. In 8.41, she escorted the first Russian convoy as part of Operation 'DERVISH'.

After a refit on the Humber 22.9-26.11.41, INGLEFIELD had seven arduous months with her Flotilla. The highlights of this period of service were participation in the flying-off operation to Malta with the US carrier WASP on 20.4.42, in addition covering elements of the Home Fleet for convoys PQ15 and QP11 and escort duties with QP12 from Kola between 21-31.5.42.

On the reorganisation of the Home Fleet Flotillas in 6.42, INGLEFIELD joined the 8th D.F. where she remained for next 9 months. That same month she and her sister INTREPID made passage to the Kola Inlet 10-14.6.42 with spare parts and ammunition for warships at Kola, returning as part of the escort for convoy QP13 to Iceland, where she arrived on 5.7.42. After refitting on the Humber 31.8-8.11.42 INGLEFIELD spent the winter of 1942-43 escorting the Russian convoys—JW51A, RA51, JW52, RA52, JW53 and finally RA53 which sailed from Kola on 1.3.43.

In 3.43, with the crisis on the North Atlantic at its height, INGLEFIELD was one of the vessels attached to the Western Approaches Command and served with the 4th Escort Group for two months before rejoining her Flotilla on 15.5.43. Almost immediately INGLEFIELD escorted the battleship HOWE to Gibraltar, returning as escort for the battleship NELSON. She was, however, not to remain long in the UK as the next month she was one of the reinforcements for the Mediterranean Fleet for the invasion of Sicily on 10.7.43. There she performed bombardment duties for several days, before participating in the sinking of the Italian submarine ASCIANGHI on 23.7.43 off the invasion beaches. INGLEFIELD then took part in the invasion of Italy, escorted the Italian Fleet to Malta and thereafter patrolled the Straits of Otranto.

INGLEFIELD then briefly returned to the UK during 10.43, before operating in the Bay of Biscay the next month and returned to the Mediterranean before the end of 11.43. After a period under repair at Gibraltar she provided gun support for the invasion of Anzio 22.1-25.2.44. On the latter day when three miles off Anzio light, the distinguished career of INGLEFIELD came to an end when she was hit and sunk by a German glider bomb. 35 of her crew died. There were 157 survivors. She lies in 41°26'N 12°38'E.

ICARUS, seen here in 7.37, spent 16 months on minelaying duties until 4.41 and then undertook fleet duties in the Arctic and Mediterranean before spending the last two years of the war on Atlantic escort duties.

ICARUS (D03)

ICARUS on commissioning was attached to the Home Fleet before joining the 3rd D.F. on 12.7.37 and the Mediterranean Fleet. On her arrival at Gibraltar she escorted the crippled destroyer HUNTER from Gibraltar to Malta. ICARUS served with the Mediterranean Fleet until 9.39 and during this time she participated in an Aegean cruise between 8-11.37 and operated off the Spanish coast during 4.38 and again between 8-11.38. Between 28.11.38-14.2.39 she refitted at Malta incorporating the works necessary for minelaying duties. The last months of peace were spent in the eastern Mediterranean. On 1.9.39 ICARUS collided with the Greek vessel MICHALIS off Alexandria and after temporary repairs at Alexandria by the repair ship RESOURCE, she sailed for Malta, where permanent repairs were effected between 12.9-6.10.39. On completion of the repairs she sailed for the U.K. and arrived at Plymouth five days later. She finally joined the 3rd D.F. at Scapa on 9.11.39. On 29.11.39 she had an early success, when with the destroyers KINGSTON and KASHMIR, she sank U35 off the Shetland Islands. On 16.12.39 she rescued 13 survivors from the Norwegian ship GLITREFJELL, mined on passage from Immingham to Stockholm.

During 12.39, ICARUS joined EXPRESS, ESK, IVANHOE, INTREPID and IMPULSIVE in the newly formed 20th Minelaying Flotilla, based at Immingham. After refitting and completing her conversion to a minelayer at Portsmouth between 24.1-26.2.40, ICARUS undertook minelaying operations in the Heligoland Bight and the Moray Firth. On 5.4.40 she left Scapa to lay a minefield in Vest Fjord at the start of the Norwegian campaign. On 10.4.40 she sank the 3,800 ton iron-ore ship EUROPA and the next day off the Lofoten islands captured the 8,514 ton ALSTER which was carrying over eight thousand tons of stores for the German Army.

On 13.4.40, ICARUS was one of the escorts for the battleship WARSPITE during the destruction of eight German destroyers and a U-boat at the Second Battle of Narvik. Other operations undertaken included the landing of allied troops at Andalsnes on 23.4.40 and a minelaying operation off Trondheim on the night of 29/30.4.40.

Following a boiler clean at Immingham between 7-21.5.40, ICARUS was ordered to Dover to participate in Operation 'DYNAMO' and in the four days before 1.6.40 when accumulated damage necessitated her withdrawal, ICARUS safely evacuated 4704 soldiers.

Repairs were completed at Portsmouth on 18.6.40, and she was also fitted with a 12 pounder gun. ICARUS then returned to her duties with the 20th D.F. until its disbandment during 4.41. She conducted "lays" in the North Sea, operated from Portsmouth during 11.40, laid mines off Brest during 2.41 and later that same month (26.2.41) she laid the first British magnetic moored mines off Cap d'Antifer. Her last lay was carried out on 20.4.41 off western France.

ICARUS had been briefly refitted at the yard of Brigham and Cowan on the Humber between 11.1-7.2.41, but this was followed by a further refit between 23.4-14.5.41, before she worked up at Scapa as a fleet destroyer for duties with the Home Fleet 3rd D.F.

ICARUS was soon in action, as she was one of the destroyers detailed to escort the battle-cruiser HOOD and battleship PRINCE OF WALES, attempting to find the German battleship BISMARCK. After HOOD was blown up in the action of 24.5.41 ICARUS searched for survivors—only three were found. On her return to Scapa ICARUS escorted the newly completed aircraft carrier VICTORIOUS.

She then operated with the 3rd D.F. for the next year and participated in the Spitzbergen raid of 18.8.41 and covered convoys PQ7B and QP5 to and from Russia during 1.42. Earlier she had escorted the damaged battleship NELSON (torpedoed during the passage of the 'HALBERD' convoy to Malta) to Rosyth, where they arrived 23.11.41. Other operations during the first months of 1942 included a sweep of the Norwegian Coast on 23.2.42, an attempted interception of the battleship TIRPITZ on 8.3.42 and the unsuccessful attempt to escort the damaged cruiser TRINIDAD to the U.K. from Murmansk 13/14.5.42. During this period she repaired a damaged port propeller at Ardrossan 2-11.7.41, refitted at the yard of Amos & Smith at Hull between 11.9-14.10.41 and had a further refit at Rosyth 12.4-2.5.42.

On the reorganisation of the Home Fleet destroyer flotillas during 6.42 ICARUS joined the 8th D.F. She was engaged in the 'HARPOON' convoy operation to Malta and although heavily attacked she escaped undamaged and left Gibraltar on 18.6.42 as one of the escorts for the aircraft carrier ARGUS returning to the U.K. ICARUS then acted as escort for the Home Fleet covering the ill-fated convoy PQ17 and returned to Scapa on 8.7.42. Between 10-15.8.42 she participated as an escort in the 'PEDESTAL' convoy operation to Malta. ICARUS picked up 2 officers and 20 ratings from the torpedoed aircraft carrier EAGLE on 11.8.42.

On her return to the U.K. ICARUS refitted at the yard of Amos & Smith between 5.9-16.10.42. She then returned to her duties with the 8th D.F. of the Home Fleet in the Arctic; escorting convoys QP15, JW51B, RA52, JW53 and RA53 between 11.42-3.43. In 4.43, she was one of the Home Fleet destroyers loaned to Western Approaches Command at the height of the crisis in the North Atlantic and escorted convoy HX232. However ICARUS was in dire need of another refit, which was undertaken at the Humber Graving Dock between 18.4-30.6.43. At the same time she was converted into an anti-submarine escort destroyer and re-armed (*see* Appendix II).

On 21.6.43, ICARUS transferred to Escort Group C2, based at Londonderry and for the next eleven months undertook convoy escort duties. On 5.3.44, whilst supporting convoy HX280, ICARUS detected a shadowing U-boat and after a 30 hour hunt, U744 was forced to the surface and sunk, some 450 miles west of Valentia—28 of her crew were picked up.

ICARUS late in the war with many modifications, including augmented anti-aircraft armament, an increased depth charge capacity and the fitting of radar. Note also the carley floats beside the bridge.
(National Maritime Museum N31743)

During 5.44, ICARUS was transferred to the 14th Escort Group and undertook escort duties in the Channel until 28.7.44, when she arrived at Wallsend to refit at the yard of Wallsend Slipway. This was completed by 16.10.44. She then worked up at Tobermory and spent the remainder of the war on local escort duties with the 14th Escort Group in the Irish Sea and Channel. On 22.1.45 ICARUS participated in the sinking of her third U-boat U1199 off Falmouth, by depth charge and Hedgehog.

In 6.45, ICARUS relieved the sloop BRIDGEWATER for duties with the 3rd Submarine Flotilla on the Clyde until 11.45. ICARUS then participated in Operation 'DEADLIGHT', the scuttling of German U-boats in the Atlantic between 11-12.45. On 26.1.46 she sailed for Libau as escort for the submarine U3515, arriving on 2.2.46 when the submarine was transferred to the U.S.S.R. She returned to Portsmouth on 7.2.46. She then returned to her duties on the Clyde until 7.46 when she reduced to Category 'C' Reserve. ICARUS arrived at the yard of the West of Scotland Shipbreaking Co. Ltd. at Troon on 29.10.46.

ILEX (D61)

ILEX joined the 3rd D.F. on 21.7.37 and six days later sailed for service with the Mediterranean Fleet. She arrived at Malta on 3.8.37 in time to participate in the Autumn cruise to the Aegean and the Dardanelles that ended in 11.37. ILEX then remained at Malta for a month before undertaking patrols based at Oran and Gibraltar until 1.38. She then undertook peace-time duties until the outbreak of war, punctuated by a refit at Malta between 14.10.38-10.1.39.

On the outbreak of war her flotilla returned to Plymouth for duties with the Western Approaches Command. Success came quickly because on 13.10.39 in company with IMOGEN she sank U42 some 290 miles west of Fastnet.

ILEX and her sisters transferred to the Home Fleet at the beginning of 11.39 and operated with that fleet until recalled to the Mediterranean on 17.5.40. During this period, she was under repair at Liverpool 3.1-2.2.40. On 5.4.40, ILEX with ISIS, IMOGEN and her leader INGLEFIELD escorted the minelayer TEVIOTBANK to lay a minefield off Stadtlandet, Norway. On the invasion of Norway, ILEX remained in the Namsos and Tromso area for several weeks and did not arrive on the Clyde until 29.4.40 to escort the transports BELLEROPHON and LYCAEON to Narvik.

ILEX was then ordered to the Mediterranean, leaving Plymouth on 17.5.40 and arriving with several other destroyers at Alexandria eleven days later to join the 2nd D.F. ILEX was to participate in many actions during the next year—the action off Calabria on 9.7.40, the Australian cruiser SYDNEY's sinking of the Italian cruiser BARTOLOMEO COLLEONI off Cape Spada on 19.7.40—where ILEX rescued 230 survivors. She escorted the aircraft carrier ILLUSTRIOUS for the strike at Taranto on 11.11.40 and participated in the Cape Matapan action of 26.3.41.

ILEX was heavily engaged in the Greek and Cretan campaigns, bombarding the airfield at Scarpanto on the night of 20/21.5.41. Two days later her propeller was damaged by a near miss and she was under repair for 10 days at Alexandria. On 15.6.41, whilst screening a division of British cruisers between Beirut and Haifa, she was attacked by German dive bombers, which released two bombs simultaneously, one of which landed on the starboard side 15ft from the vessel and the other 50ft on the port side. The first explosion resulted in a low frequency whip of the entire vessel, which shook the foremast so violently that the radar aerial sheared. Steam pressure was lost partially due to damage and a misunderstanding on the part of the crew. The Captain later closed down No.2 boiler, which had been re-started, because of contaminated fuel. She was towed into Haifa where examination showed a general transverse corrugation of the starboard shell plating from frames 60 to 98. The hull was also buckled at the midship region, by No.3 boiler room, taking the form of a wrinkle that girded the hull between frames 89 and 93 up to a depth of 8 inches. At Haifa temporary repairs were undertaken that included the welding of angles along the keel. She left Haifa in tow for Port Said on 29.6.41. At Suez additional repairs were undertaken, which included the fitting of longitudinals between bulkheads 86 and 98, spaced 18 inches apart on either side of the vertical keel.

On 20.7.41 she left Port Said and arrived fours days later at Aden, where she remained until 15.9.41 whilst efforts were made to stem the leakage in No.3 boiler room. Her passage around Africa was slow, being further delayed by engine trouble at Mombasa, not reaching Durban until 11.41. Further strengthening took place there in order to reduce the hog of the ship. She remained at Durban until 5.2.42, when she sailed via the Cape, Freetown, Pernambuco, Trinidad to Charleston, U.S.A. where she finally arrived on 14.3.42 to refit. A maintenance crew of one officer, one gunner and some 30 ratings was retained.

A superb photograph of ILEX in 9.42 on completion of her refit at Charleston Navy Yard, following damage off the Lebanese coast on 15.6.41. Note the U.S. Motorboat. (U.S. Navy)

ILEX's repairs were not completed until 9.42 and after working up at Key West with a new crew until 10.42, she escorted convoy WS24 from Bahia to the Cape. After being detached from this convoy, ILEX proceeded to Freetown, where she was stationed until 1.2.43. She was then allocated to the 6th D.F. in the Mediterranean and on 3.6.43 she carried out a bombardment of Pantellaria with her sister ISIS.

The next month ILEX was transferred to the 8th D.F. and participated in the Sicily landings. Three days later on 13.7.43 with ECHO, she sank the Italian submarine NEREIDE south-east of Messina Straits. ILEX then participated in the invasion of Salerno on 15.9.43 and was thereafter deployed on routine duties until the close of the year.

In 12.43, the C. in C. Mediterranean reported that ILEX was in urgent need of a major refit. ILEX then lay at Bizerta between 17.1-22.6.44. It is not known whether her laying-up was due to her poor material condition or if she was a victim of the manpower crisis then occurring in the Royal Navy. On 30.6.44 she was listed as being laid up in reserve at Ferryville, where repairs to her superheaters were being undertaken. By 2.45 she was to be repaired at a low priority. This was again changed and by 20.3.45 she had been towed to Malta and was relegated to Category 'C' Reserve, where she remained until 6.47.

ILEX's fate is somewhat obscure but she is reported to have been sold to a Sicilian firm of breakers and broken up during late 1947.

IMOGEN in 1937 with the tricolour painted on B gun mounting. She was to survive barely three years before being lost in collision with the cruiser GLASGOW.

IMOGEN (D44)

IMOGEN's career was to last almost exactly three years, from 12.7.37 when she joined the 3rd D.F. to the day she was lost on 16.7.40. All her service was with the 3rd D.F. After fitting of the Mountbatten station-keeping gear and trials, she worked up at Portland, before leaving for Gibraltar on 10.8.37. She then left Gibraltar as escort for HUNTER on passage to Malta to complete her mine damage repairs. After participating in the Mediterranean Fleet cruise to the Aegean, she operated from Malta between 1.11-20.12.37. During 1938 she patrolled off the Spanish Mediterranean coast during 6-7.38 and 9-10.38, escorting the s.s. AFRICAN TRADER to territorial waters at the entrance to Valencia on 16.6.38. She refitted at Malta between 17.10.38-28.11.38.

IMOGEN then operated on normal peace-time duties with the Mediterranean Fleet until recalled to the U.K. for refitting. However, she only completed a docking at Sheerness between 3-26.8.39, when she left for the Mediterranean. She arrived at Malta on 3.9.39, but sailed again two days later and arrived at Plymouth on 11.9.39 for duty with the Western Approaches Command. The next month, the 3rd D.F. was transferred to the Home Fleet.

On 13.10.39 IMOGEN and ILEX sank U42 in postion 49° 12'N 16° 00'W after the latter had attempted to torpedo the merchant ship STONEPOOL which IMOGEN then escorted to Barry. On passage she rescued the survivors of the vessels LOUISIANE and BRETAGNE sunk by U48 and U45 some 130 miles off Fasnet on the 13th and 14th respectively.

IMOGEN then repaired at Liverpool 20.10-7.11.39 before operating on patrol and escort duties with the Home Fleet until 4.40. On 28.12.39, she assisted the battleship BARHAM, torpedoed by U30 in position 58° 47'N 08° 05'W, off the Butt of Lewis. Another highlight was the sinking of U65 by IMOGEN, INGLEFIELD, ESCORT and the submarine NARWHAL about 90 miles east of Orkney on 25.2.40. IMOGEN then operated with the Home Fleet in Norwegian waters during 4-5.40 and arrived at Plymouth on 18.6.40 to escort the newly commissioned aircraft carrier ILLUSTRIOUS on the most dangerous portion of her passage to Bermuda for work up.

At about midnight on 16.7.40 when returning to Scapa from an operation, she was in collision with the cruiser GLASGOW in thick fog off Duncansby Head. IMOGEN was extensively damaged, caught fire and was abandoned, sinking in position 58°34'N 02°54'W. 19 ratings were lost but 10 officers and 125 ratings were picked up by GLASGOW and landed at Scapa.

IMPERIAL (D09)

After working up, IMPERIAL joined the 3rd D.F. on 10.7.37. However, her departure to the Mediterranean was delayed by the late fitting of her gun mountings. She did not finally reach Malta until 17.9.37, after being further delayed by defects with her boiler brickwork. IMPERIAL then undertook normal peace-time duties until taken in hand for repairs at Malta 24.2-2.3.38 to her forecatle deck beams and other defects. She then undertook patrol duties off Barcelona and Palma during 6-7.38.

After a further refit at Malta between 14.10-5.12.38, IMPERIAL resumed her duties with the 3rd D.F. until the outbreak of war and was lying at Alexandria on 3.9.39. She then left Malta two days later to join the Western Approaches Command with her flotilla and was based at Plymouth for two months before being transferred to the Home Fleet during 11.39.

IMPERIAL—a fine pre war view. She was to be one of six British destroyers, including her near sisters GREYHOUND and HEREWARD, lost off Crete in 5.41.

IMPERIAL then operated on local escort duties with the Home Fleet throughout the bitter winter of 1939/40, suffering a variety of small defects. However on 26.2.40 she was in collision with and sank the Swedish freighter NORDIA in position 61°12'N 03°08'E. IMPERIAL was flooded to No. 7 bulkhead and was unable to steam at more than 12 knots and was escorted towards Lerwick by the AA cruiser CALCUTTA. IMPERIAL's damaged bow plating was repaired at Swan Hunter and Wigham Richardson's yard on the Tyne 5.3-14.4.40. She then participated in some of the later actions of the Norwegian Campaign and on 3.5.40 escorted the cruisers MONTCALM (French) and DEVONSHIRE, covering the withdrawal of allied forces from Namsos.

IMPERIAL was one of the Home Fleet reinforcements that left Plymouth for the Mediterranean on 15.5.40. She arrived at Alexandria on 24.5.40 and joined the 2nd D.F. She was to serve with this flotilla until her loss a year later. She participated in many actions, including Operation "HATS" — the replenishment convoy for the Mediterranean Fleet between 29.8-6.9.40. On 11.10.40, when participating in convoy operation MB6, IMPERIAL was mined 15 miles south of Delimara in position 35°24'N 14°34'E. She suffered considerable damage but was docked at Malta the same day. She was under repair until 23.3.41 which included the fitting of GALLANT's Hastie steering gear. She immediately left for Gibraltar, where she worked up, before returning to the Eastern Mediterranean.

She was not to survive long. On 29.5.41, whilst under German aircraft attack during the evacuation of Crete, she was near missed which damaged her steering, without her officers being aware of the damage. Some hours later when approaching the Kaso Strait, IMPERIAL suddenly started to turn in circles. Rear Admiral Rawlings, the Commander of the Force, realised that the ship was too close to Crete to be towed home. Under cover of darkness, the troops and crew of the IMPERIAL were transferred to HOTSPUR, which then sank her with two torpedoes. IMPERIAL lies in 35° 23' N 25° 40' E.

IMPULSIVE (D11)

IMPULSIVE undertook torpedo discharge trials at Portland between 14-16.2.38, before joining the 3rd D.F. at Malta on 24.2.38. She spent the next 18 months in the Mediterranean, participating in the Mediterranean Fleet's summer cruise between 6-8.38 as well as duty off the Spanish coast during 4.38 and 3.39. During the Munich crisis, she was one of the escorts between 28-30.9.38 for the battle-cruiser HOOD and the liner AQUITANIA until the latter was far out into the Atlantic.

She refitted and converted to a minelayer at Malta between 28.11.38-20.2.39, serving in the Aegean off Crete during 4-5.39 and off Palestine during 6.39, before operating between Cyprus and Alexandria. On 3.9.39, IMPULSIVE was on passage between Alexandria and Malta.

The 3rd Flotilla immediately returned to the U.K., with IMPULSIVE arriving at Plymouth on 14.9.39, with the survivors of the sunken aircraft carrier COURAGEOUS. After some six weeks with the Western Approaches Command, the 3rd D.F. joined the Home Fleet at Scapa on 25.10.39. During her three months with the Home Fleet the principal incident was IMPULSIVE's towing of the destroyer leader DUNCAN into Invergordon following collision damage.

Between 24-26.1.40, IMPULSIVE refitted and converted to a minelayer at Portsmouth, before joining the 20th D.F. (M.L.) at Immingham the next day. She was to serve with this flotilla for the next few months, before being attached to the 21st D.F. in 9.40 for two months. IMPULSIVE participated in a minelaying operation in the Heligoland Bight during 3.40 and the mining operation off Vest Fjord on 8.4.40 at the time of the German invasion of Norway. This was followed by Operation 'SICKLE' — the landing of the troops at Molde and Andalsnes on 23.4.40 and a minelaying operation off Trondheim on the night of 29.4.40.

IMPULSIVE as a minelayer, with A and Y gun mountings landed and carrying a full complement of mines. Note the minelaying sponsons aft.
(Courtesy R.L. Wilson)

Further operations off Norway then followed, before IMPULSIVE went south to participate in the Dunkirk evacuation on 29.5.40. She was credited with the evacuation of 2,919 troops, before she damaged a propeller on wreckage and was withdrawn for repairs at the Blackwall yard of Green & Silley Weir between 3.6-10.7.40. IMPULSIVE then returned to her minelaying duties but was boiler cleaning on the Humber between 25.8-3.9.40 and hence missed the disaster off Texel on the night of 31.8/1.9.40.

IMPULSIVE refitted at the Hull yard of Brigham and Cowan 7.11.40-18.1.41. After re-converting to a minelayer between 24-28.1.41, she operated with the 20th D.F. until 5.41. She carried out her last lay off the Faroes on 5.5.41, before the flotilla was disbanded. IMPULSIVE briefly operated with the Home Fleet during 4.5.41 and during this period she rescued 278 survivors from the A.M.C. SALOPIAN torpedoed by U98 on 13.5.41, and conveyed them to Clyde, where they were landed on 19.5.41.

On 28.5.41 IMPULSIVE reported serious damage to No.3 boiler room and repairs on the Humber by Amos & Smith were completed between 18.6-6.8.41.

She then served with the Home Fleet's 3rd D.F. for the next four months, participating in the escort of convoy PQ1 between Hvalfiord and Archangel between 29.9-11.10.41 and with ESCAPADE escorted the cruiser SUFFOLK from Archangel to the U.K. between 19-27.10.41. However because of accumulated defects, IMPULSIVE again refitted, this time at Thornycroft's yard at Southampton between 30.12.41-27.7.42.

After working up, she served with the Home Fleet's 8th D.F. for the next five months. Between 2-16.9.42, she escorted convoy PQ18 to Russia, sinking U457 without survivors on the latter day. Subsequently she escorted convoys QP14 and QP15 and provided distant cover for JW51B, as escort for the battleship ANSON.

IMPULSIVE was one of the Home Fleet destroyers loaned to the Western Approaches Command during the convoy crisis of 3-5.43 and was attached to the 3rd Support Group at this time. She again refitted at the yard of Amos & Smith at Hull 6.43-22.7.43 and then operated with the Home Fleet as an unattached vessel until 5.44. She participated in diversionary duties off Norway on 10.7.43 and some other minor operations before covering convoy JW54A from Loch Ewe to Kola between 15-24.11.43, returning with convoy RA54B, which left Archangel on 26.11.43. IMPULSIVE was one of the escorts for convoy JW55 that was threatend by the German battle-cruiser SCHARNHORST, before the latter was sunk on 26.12.43.

During 2.44 IMPULSIVE participated in an anti-shipping sweep off the Norwegian coast — Operation 'POSTHORN' — before escorting convoys RA57, JW58 and RA58 during 3-4.44. On 25.2.44, IMPULSIVE rescued the few survivors of the destroyer MAHRATTA torpedoed by U990 in the Barents Sea.

After her transfer to the 23rd D.F. during 4.44, IMPULSIVE was allocated to shore support duties off ''SWORD'' beach on 'D' Day. She undertook these duties until she refitted at Immingham by the Humber Graving Dock Ltd, between 26.7-23.9.44.

IMPULSIVE was then temporarily attached to the 14th Escort Group of the Western Approaches Command on Atlantic escort duties, before being transferred to Portsmouth Command's 8th D.F. during 3.45 for the last months of the war in the Channel.

At the end of hostilities, IMPULSIVE visited St. Peter Port, Guernsey on 14.5.45 and Jersey on 7.6.45 as escort for the cruiser JAMAICA which was conveying H.M. The King on his visit to Channel Islands.

On 17.6.45, IMPULSIVE entered Category 'B' Reserve at Harwich and reduced to Category 'C' Reserve on 13.3.46, pending her disposal by scrapping. She arrived at the yard of Thomas Young and Sons Ltd. at Sunderland on 22.1.46 for demolition.

INTREPID on builders' trials during 6.37 without her main armament. Delays were already being experienced in the delivery of gun mountings.
(National Maritime Museum N3100)

INTREPID (D10)

After commissioning INTREPID worked up at Portland before joining the 3rd D.F. in the Mediterranean until the outbreak of war. During 12.37-1.3.38, she participated in combined exercises with the French Fleet at Oran, followed by three months stationed at Malta. By 3.38 she was operating off the south coast of Spain — this being the first of four occasions that she was to operate in that area. The other periods of service were during 8.38, 10.38 and 2-3.39. Earlier, between 19-26.4.38 INTREPID was quarantined at Malta with a suspected case of typhus. A pleasant interlude was her participation in the Mediterranean Fleet cruise in the Aegean and to Yugoslavia in 7.38. Two months later, INTREPID acted as escort for the battle-cruiser HOOD and the liner AQUITANIA during the Munich crisis.

Between 2.11.38-23.2.39 at Malta INTREPID refitted and completed the necessary changes to convert her into a minelayer. She then operated from Malta between 4-5.39, Alexandria 5-6.39 and Cyprus and Palestine during 7.39.

On 3.9.39 INTREPID was on passage between Alexandria and Malta, with orders to return to the U.K., where she arrived on 15.9.39. She was immediately embroiled in the war, as three days later she was searching for survivors from the torpedoed aircraft carrier COURAGEOUS. INTREPID was then attached to the 5th D.F. of the Home Fleet at Scapa until 6.11.39. After refitting at Sheerness between 14.11-13.12.39, she returned to her duties in the Channel for the next month.

On the formation of the 20th Destroyer Minelaying Flotilla during 1.40, she converted to a minelayer at Sheerness between 20-25.1.40 and was to serve with this flotilla until 4.41. On the night of 17/18.2.40 she participated in the ALTMARK operation. A month later on 17.3.40 she collided with and sank the trawler OCEAN DRIFT, picking up eight survivors and returned them to Invergordon. After repairs to her bows undertaken by Smith's Dock at Middlesbrough between 22.3-28.4.40, INTREPID returned to her minelaying duties until 28.5.40.

On this day, she joined Operation 'DYNAMO'. She was damaged by aircraft on 30.5.40 and repairs at Middlesbrough were completed on 12.6.40.

She then returned to her minelaying duties and survived unscathed the disaster of 31.8 off Texel. However, on 1.11.40 a mine exploded on her starboard quarter in position 54°30′N 00°53′ W. Her engines were put out of action and INTREPID was towed into Hartlepool, where repairs were completed between 2.11-20.12.40. Until her departure for Scapa 26.4.41, she operated from Immingham, with detachments to Dartmouth between 22.2-25.3.41 and Plymouth between 25.3-23.4.41.

INTREPID then operated with the Home Fleet's 3rd D.F. until 3.42, undergoing general repairs at Humber Graving Dock, Immingham between 5.6-15.6.41 and a refit at Green & Silley Weir at Tilbury between 29.8-22.10.41. She had participated in the early stages of the BISMARCK sortie, before having to return to Iceland for fuel. Between 22.7-4.8.41 she participated in the British aircraft carrier raid at Kirkenes and Petsamo and later (24.11.41) with the cruiser KENYA and the destroyer BEDOUIN and two Russian destroyers, unsuccessfully searched for German vessels off the northern polar coast between Nordkyn and Vardo.

INTREPID spent the remainder of the winter of 1941/42 on escort duties, protecting convoys PQ3, QP3, PQ9, PQ10, QP7, PQ12, QP8 to and from north Russia, until she refitted at Doig's Grimsby yard 24.3-30.4.42. On completion of this refit, INTREPID participated in Operation 'BOWERY' by escorting the aircraft carriers EAGLE and U.S.S. WASP flying off aircraft to Malta on 9.5.42. She then returned to Arctic escort duties with convoy QP13 to Iceland from North Russia at the time of the disastrous PQ17 convoy.

A fine action shot of the largely unaltered INTREPID in early 1942 when a member of the Home Fleet's 3rd D.F.

Her busy career continued with the vessel returning to the Mediterranean for Operation 'PEDESTAL' — the hard fought and successful convoy action to Malta between 10-15.8.42. On 11.8.42, she picked up two officers and 20 ratings from the torpedoed aircraft carrier EAGLE.

The next four months were to see INTREPID heavily involved in Arctic convoys, being one of 16 destroyers of the close escort of PQ18 from Iceland to North Russia 12-18.9.42. She picked up survivors from the destroyer SOMALI torpedoed on 20.9, which sank in tow on 24.9.42. Between 13-28.10.42, the cruiser ARGONAUT with INTREPID and OBDURATE made a return passage to Murmansk to take hospital equipment to Russia, returning with survivors from PQ18.

INTREPID had a much needed refit on the Humber between 2.12.42-26.1.43, before resuming her duties with the 8th D.F., in which she had served since 5.42. Between 19.2-14.3.43, INTREPID undertook the escort of her last Arctic convoys — JW53 and RA53. She was then attached to the 4th Escort Group at Greenock as a reinforcement during the peak of the Atlantic U-Boat campaign until 5.43.

During 6.43, the 8th D.F. was sent as a reinforcement to the Mediterranean Fleet and participated in the invasion of Sicily on 10.7.43, when based at Malta. On 10.9.43 INTREPID was part of the escort for the surrendered Italian fleet to Malta. She transferred to the Eastern Mediterranean at Alexandria on 16.9.43.

After participating in a trooping operation to Leros on 20-21.9.43, she was lying at Leros Harbour at 09.00 hours on 26.9.43, when the island was raided by Ju88's of LG1. During the raid the Greek destroyer VASILISSA OLGA was sunk and a hole 6ft x 3ft above the boiler room was suffered by INTREPID. The main port steam line from Nos. 2 and 3 boilers was perforated, causing the ship to lose all steam. She was moved to the eastern part of Leros Harbour, but at approximately 16.47 hours she was hit aft and her stern blown off up to 'X' gun. Her captain was pinned down between the after superstructure and the starboard winch. After he had been released the ship was abandoned. The crew later returned to retrieve stores and at 21.00 hours the ship was inspected but the Engineering Officer considered that she could sink at any moment as previously plugged holes in the engine room were leaking. The party was withdrawn and INTREPID capsized and sank at about 02.00 hours on 27.9.43. 15 of her crew were killed.

ISIS (D87)

ISIS worked up at Plymouth, before joining the Mediterranean Fleet's 3rd D.F. On 20.9.37 she touched ground at the entrance to Mudros Harbour and was taken in hand four days later for docking and repairs, which were not completed until 28.10.37. Her normal peacetime duties were interrupted during the Munich crisis, when she escorted the liner AQUITANIA, with the battle-cruiser HOOD, from Gibraltar well out into the Atlantic returning to Gibraltar on 1.10.38. She then refitted at Malta between 14.10.38-10.1.39.

On 7.9.39 she left Gibraltar for Plymouth, where she joined the Western Approaches Command four days later. Until 2.40, ISIS was engaged on a variety of duties with the 3rd D.F., the most conspicuous being the capture in the Atlantic of the German steamship LEANDER (990 tons), which was sent into Falmouth on 9.11.39. Six days later, off the Hebrides, she tried to tow the derelict tanker ARNE KJODE (torpedoed by U41 on 12.11.39 in position 58°51'N 08°07'W) but eventually had to sink her.

ISIS re-tubed her superheaters at Falmouth between 8.2-12.3.40, before being embroiled in the Norwegian Campaign. However, on 8.5.40 she struck a submerged wreck in Balangen Fiord and wrecked her propellers before they could be stopped. Her sister ship ILEX towed ISIS into Skjael Fiord and she was subsequently towed to Greenock, where she arrived 17.5.40. The tow was continued by the tug BUCCANEER escorted by the destroyer WITCH to Falmouth, where temporary repairs were completed on 13.6.40. Permanent underwater damage repairs were completed at Devonport between 14.6-20.9.40.

ISIS then worked up at Scapa and after undertaking some local escort work to the Faroes, she left on 31.10.40 to join the 13th D.F. at Gibraltar arriving six days later. She then undertook convoy duty between Gibraltar and Freetown until 2.41. This was punctuated by attachment to Force 'H' for operations 'WHITE,' 'COLLAR' and 'COAT' — reinforcement convoys to Malta and by her participation in the bombardment of Genoa.

She then circumnavigated the Cape as a convoy escort, before joining the Mediterranean Fleet at Alexandria on 9.4.41 in time to participate in the disastrous Greek and Cretan campaigns. On 9.6.41, ISIS and HOTSPUR forced the French destroyers GUEPARD and VALMY to retire from attacking the disabled JANUS off the Lebanese Coast.

On 15.6.41 in company of the cruiser PHOEBE and the destroyer DEFENDER, the vessels were attacked by Ju88's of II/LG1. At 17.10 hours ISIS was turning to starboard at full speed, when she was twice near-missed on her starboard side by No. 60 station and 33 station by 500kg bombs. The vessel whipped violently and was lifted some 6'. A 10' long by 2' wide fracture occurred on the starboard side at station 60. There was also considerable distortion and buckling of the decks and the main hull structure at station 134. The boiler room was flooded and the vessel listed 5° to starboard. Her maximum speed was 10 knots with excessive trim by the bows. The radio and anti-submarine systems were out of order, although her 4.7" guns could be fired by local control.

ISIS down at the head and listing after being badly damaged by near misses off the Syrian coast on 15.6.41. She suffered shock and whip damage and was out of action until 11.43.

After temporary repairs at Haifa, ISIS left there on 28.7.41 for Alexandria on passage to Singapore, where she finally arrived on 13.10.41 after calls at Port Said, Aden and Bombay en route. She was immediately taken in hand for repairs, which were scheduled to complete on 21.3.42. However, ISIS was further damaged by an air raid on Singapore on 17.1.42 and it was decided that she could be towed to Batavia, where she arrived on 7.2.42. The rapidly deteriorating situation meant that a further move was needed and she was towed via Trincomalee to Bombay, where repairs were again started on 4.3.42. The much delayed repairs were not finally completed until 17.1.43, when she commissioned for service with the 12th D.F. of the Mediterranean Fleet.

On 19.2.43, whilst escorting convoy XT3, north-east of Benghazi, ISIS participated in the sinking of U562 in position 32°57'N 20°54'E. During 4.43 she transferred to the 14th D.F. and remained in the Mediterranean until 5.44. The final three months of her service were spent with the 8th D.F. The highlights of her career were the bombardment of Pantellaria during 6.43, her participation in the invasion of Sicily on 10.7.43 and the escorting of surrendered Italian submarines from Bone to Malta, where they arrived on 17.9.43.

On 12.5.44, the re-allocated 12th D.F. arrived at Scapa for training with the Home Fleet for invasion duties, which were performed during 6-7.44. However on 20.7.44, whilst in channel 'T' off the Western sector of the Normandy landing area, ISIS detonated a mine and sank quickly with the loss of 11 officers and 143 ratings — the last British A/I to be lost in action. She lies in position 49°27'N 00°41'W.

A pre-war view of IVANHOE, showing a range clock, which was probably used for the calibration of her guns. The author's father was rescued by this vessel at Dunkirk.

IVANHOE (D16)

After fitting Mountbatten Station-keeping gear, IVANHOE joined the 3rd D.F. in the Mediterranean station during 9.37. She remained on station until the outbreak of war. Between 12.37-1.38, she participated in exercises with the French fleet, based at Oran. The exercises ended prematurely for IVANHOE as she was suffering from buckled tubes in her superheaters which required the partial re-tubing of her boilers at Malta between 15.1-19.3.38. She then operated off the Spanish coast during 3-4.38, based at Gibraltar and three months later was operating off Palma and Valencia. She was at Cartagena in 2-3.39 throughout the evacuation of foreign nationals and Republican troops and supporters. On 3.9.39 IVANHOE was on passage between Alexandria and Malta and by 14.9.39 was lying at Plymouth as part of the Western Approaches Command to which the 3rd Flotilla had been transferred. Four days later she was in action with U29 which tried to torpedo her.

In 10.39 IVANHOE was briefly attached to the 5th Flotilla of the Home Fleet, but then refitted at Sheerness between 14.11-13.12.39. She converted to a minelayer at the same time. On the night of the 17/18.12.39 with ESK, EXPRESS and INTREPID, IVANHOE, now part of the 20th D.F., laid a total of 240 mines in the Ems estuary. After further minelaying operations IVANHOE re-shipped her after gun mountings and torpedo tubes at Portland between 27.1-3.2.40.

She then operated as a fleet destroyer for the next two months, participating in the ALTMARK operation on 17-18.2.40. IVANHOE and GALLANT escorted INTREPID, which had damaged her bow in collision with the fishing vessel OCEAN DRIFT, into Invergordon on 18.3.40.

IVANHOE again converted to a minelayer at Sheerness and participated in two mining operations, on 8.4.40 with ESK, ICARUS and IMPULSIVE when they laid a field off Bodo, and on 10.4.40 when with the minelayer PRINCESS VICTORIA and ESK, EXPRESS and INTREPID she laid 236 mines off Egmond. After a boiler clean between 7-15.5.40, IVANHOE then laid 164 mines off the Hook of Holland with ESK and EXPRESS.

Between 28.5.40 and 1.6.40, IVANHOE participated in the evacuation of Dunkirk. (The author's father who was then a corporal in the Royal Fusiliers was one of those rescued by the IVANHOE.) However, at approximately 07.30 hours on 1.6.40 when at the entrance to Dunkirk harbour she was attacked by German aircraft. Two bombs near-missed to port and starboard whilst a third detonated about 6ft above the upper deck aft of the forward funnel about 8ft to port of the middle line. No 1 and 2 boiler rooms were flooded. However No 3 boiler room was still operating and IVANHOE returned to Dover at 10 knots. 26 were killed including five soldiers and many were wounded.

Repairs to the bomb damage were undertaken at Sheerness and completed on 28.8.40. IVANHOE then converted into a minelayer for service with the 20th D.F. at Immingham between 29-31.8.40. She was lost the next day in the ill fated operation off Texel.

IVANHOE was stationed on EXPRESS's port quarter and went alongside EXPRESS after she had been mined. She picked up some men and then passed down EXPRESS's port quarter. The ships drifted apart and IVANHOE struck a mine under her bows, when about 100 yards off EXPRESS's port quarter.

By 01.45 hours on 1.9.40 steam had been raised and the ship proceeded at 7 knots stern first until 04.00 hours, when the main shaft fractured and the ship stopped. By 10.30 hours she was settling slowly and listing to port and by 13.45 hours the engine room commenced to flood. Half an hour later, IVANHOE was abandoned and vents opened to aid sinking. At 16.50 hours German aircraft bombed the ship as KELVIN came up and fired a torpedo, which struck IVANHOE under the aft superstructure and she sank in one minute in position 53° 26′ 42″ N 03° 45′ 24″ E.

THE BRAZILIAN 'H' CLASS

These vessels had been ordered in pairs from Vickers-Armstrongs on 6.12.37, J. S. White on 8.12.37 and the final pair from Thornycroft on 16.12.37 as part of a fleet replacement programme. During the Munich Crisis of 9.38, the Admiralty considered requisitioning not only these vessels, but also the two Greek destroyers of the VASILEFS GEORGIOS class also under construction. Extemporised fire control and armament arrangements were made for these vessels, so that they would have formed a homogeneous flotilla. The Brazilian vessels were not in fact taken over at that time, but were requisitioned on 5.9.39. When taken over, two of the vessels had already been launched and the others were in an advanced state of construction prior to launch.

On 4.9.39, at the request of the D.N.C., Vickers-Armstrongs supplied a list of the principal differences between the Brazilian vessels and their British sisters:
(i) The vessels were not fitted with any under water listening or signalling equipment.
(ii) The vessels were fitted with echo-sounding gear.
(iii) The ventilation trunks were constructed of steel not aluminium.
(iv) The C.O.'s Cabin was fitted with a bathroom and basins were fitted in all officers' cabins.
(v) Accommodation included more cabins and less storage space.
(vi) Shelter was provided for the navigating officer on the navigation bridge.
(vii) The galley was fitted with an oil range.
(viii) The vessels were fitted with Vickers high speed sweeps, instead of Admiralty type.
(ix) The magazines were insulated and fitted with cooling apparatus.

It was estimated that the effect on displacement was a gain of 0.32 tons. After completion, HESPERUS's main mast was cut down to 30ft and re-positioned on the searchlight platform. By 10.40, the survivors' armament consisted of three 4.7", one 3"AA, two 0.5" machine guns and four Lewis guns.

BUILDING PROGRAMME

	BUILDER	LAID DOWN	LAUNCHED	COMMISSIONED
HARVESTER (ex HANDY ex JURUA)	Vickers-Armstrongs (Barrow)	3.6.38	29. 9.39	23. 5.40
HAVANT (es JAVARY)	J.S. White	30.3.38	17. 7.39	19.12.39
HAVELOCK (ex JUTAHY)	J.S. White	31.5.38	16.10.39	10. 2.40
HESPERUS (ex HEARTY ex JURUENA)	Thornycroft	6.7.38	1. 8.39	22. 1.40
HIGHLANDER (ex JAGUARIBE)	Thornycroft	28.9.38	16.10.39	18. 3.40
HURRICANE (ex JAPARUA)	Vickers-Armstrongs (Barrow)	3.6.38	29. 9.39	21. 6.40

HARVESTER (ex HANDY ex JURUA) (H19)

JURUA was requisitioned whilst under construction and given the name HANDY, but was renamed in 1.40 to prevent confusion with the leader HARDY when signalling. After a brief work-up at Portland, HAVESTER joined the 9th D.F. and was immediately embroiled in the Dunkirk evacuation. During 'DYNAMO', HARVESTER transported some 3,200 troops to the U.K. She received machine-gun damage on 1.6.40, which required a few days repairs at Chatham.

On 10.6.40 she participated in the unsuccessful attempt to rescue the 51st Highland Division from St. Valery, returning to Portsmouth on 13.6.40. Between 19-26.6.40, she assisted in the evacuation of St. Jean de Luz and Bayonne. On 26.6.40 she arrived at Liverpool as escort for the ARANDORA STAR with some 3,000 Polish troops and air force personnel on board.

Between 7.40-8.9.40, HARVESTER operated on convoy escort duties in the Western Approaches with the 9th D.F. On 29.8.40 she arrived at Greenock with 90 survivors from the A.M.C. DUNVEGAN CASTLE, which had been torpedoed by U46 two days earlier.

However, between 8-18.9.40 she and five other destroyers of the flotilla were attached to Portsmouth Command for anti invasion duties. She then returned to Liverpool, making an unsuccessful search for the torpedoed British tanker BRITISH GENERAL on 7-8.10.40 and only just made it into Londonderry because of a shortage of fuel. On 30.10.40 she sank U32 some 135 miles west of the Bloody Foreland. Convoy escort duties continued for the remainder of the winter of 1940-41. On 5.12.40 she picked up 19 survivors from the m.v. SILVERPINE. She rescued 121 survivors from the Armed Boarding Vessel CRISPIN on 3.2.41 and four days later rescued the crew of a Whitley aircraft that had been escorting convoy SC20.

She then refitted at her builders' yard at Barrow between 18.3 and 18.4.41 and joined Force 'H' at Gibraltar and participated in the convoy operation 'TIGER' to Malta and Operation 'SPLICE' — a flying-off operation of aircraft again to Malta. Admiral Somerville regarded her as being insufficiently armed for such dangerous air dominated waters and she was quickly reallocated to the newly formed Newfoundland Escort

Force and made passage to St. John's joining the 14th Escort Group on 1.7.41. She operated with this force for the next three months, escorting the torpedoed A.M.C. CANTON into the Clyde during 8.41.

She then rejoined the Western Approaches Command as part of the Special Escort Force and the 28th E.G. HARVESTER had a hectic period of service visiting Gibraltar, Ponta Delgada and Norfolk, Virginia during 12.41. After a refit at Dundee between 30.1-16.4.42, she joined Escort Group B3 on 20.4.42 and carried out trials of the Type 251 radar during 5.42. She then operated on the North Atlantic for the remainder of the year.

HARVESTER in 5.42, shortly after becoming leader of Escort Group B3. She carries a Hedgehog in A gun position.
(Imperial War Musuem FL47042)

A further refit at Liverpool followed between 23.12.42-11.2.43 and she left Londonderry as escort for convoy ON167 four days later. She was to survive barely three more weeks.

On 10.3.43 HARVESTER was escorting convoy HX228, when it was attacked by U444. HARVESTER forced the submarine to the surface by depth charges and then rammed and sank it, taking five prisoners. HARVESTER was badly damaged and the French corvette ACONIT stood by her. The next morning HARVESTER was twice torpedoed, broke in two and sank in position 51°23N' 28°40W' at about 13.00 hours. ACONIT sank U432, the submarine responsible and then rescued HARVESTER's few survivors. Commander Tait, eight other officers and 136 ratings were lost.

HAVANT (ex JAVARY) (H32)

Originally ordered on 8.12.37 by the Brazilian Navy, her construction continued normally at J. S. White's Cowes yard until 5.9.39, when she was requisitioned by the Admiralty, whilst fitting out. Renamed HAVANT, she was completed during 12.39, and arrived at Portland on 8.1.40 to work up. On completion of her work up, she joined the 9th D.F. of the Western Approaches Command, based at Plymouth. Her first duty was to conduct an anti-submarine sweep between 4-9.2.40, with the destroyers ARDENT and WHITSHED.

However defects and the fitting of de-gaussing equipment kept HAVANT at Devonport until 3.40. It was suspected that the defects were caused by sabotage. She then rejoined her escort duties and on 7.4.40 sailed for Greenock to escort a convoy to Gibraltar. Whilst on passage to Greenock she was detached to the Home Fleet because of the German invasion of Norway. Initially, with her sister HESPERUS she screened the cruiser SUFFOLK transporting a detachment of Royal Marines to occupy the Faroes on 13.4.40. She returned to Scapa the next day and then escorted convoys to Narvik until 7.5.40. A week later she escorted the Cunard liners FRANCONIA and LANCASTRIA carrying troops to occupy Iceland, before returning to Greenock on 25.5.40.

HAVANT off Dunkirk on 1.6.40 after being damaged by aircraft bombs. Note the bomb holes on her starboard side by the searchlight. The two balls at the foremast indicate that she is not under control.
(National Maritime Museum D1675)

The evacuation of Dunkirk had commenced by this time and HAVANT arrived at Dover on 29.5.40 to participate in the operation. By 1.6.40 HAVANT had already transported over 2,300 troops to Dover. At 08.00 hours that day, after loading a further 500 troops, she left Dunkirk harbour and 40 minutes later went alongside the disabled destroyer IVANHOE and took on board all the troops and wounded from this vessel and then proceeded towards Dover under continuous air attack. HAVANT was shortly afterwards hit by two bombs in her engine-room and was further damaged by a bomb that exploded beneath her. The minesweeper SALTASH came alongside and rescued her troops and crew and, after making an attempt to tow her, sank her by gunfire at 10.15 hours. One officer and seven ratings were killed and 25 wounded. At least 25 soldiers were also killed. She lies in position 51°08'N 02°15'49"E.

HAVELOCK (ex JUTAHY) (H88)

After working up, HAVELOCK joined the 9th D.F. which was assigned to the Western Approaches Command. She was to remain with this flotilla until she became the leader of Escort Group B5 during 3.42.

HAVELOCK was immediately involved in gun support operations in the Rombaskfjord and the advance to Narvik on 27-28.5.40. However, the success was to be short-lived, as between 4-8.6.40, HAVELOCK co-ordinated the evacuation of the 24,500 soldiers from Narvik and Harstad. She finally returned to Scapa on 10.6.40.

HAVELOCK then rejoined the 9th D.F. on convoy protection duties for the next 21 months. Her duties were arduous, unrelenting and to a large extent unrewarded. A relief from her Atlantic duties was her participation in operation 'TIGER' — a joint convoy operation to supply Malta and Alexandria, between 6-12.5.41.

After becoming Flotilla leader of the Escort Group B5 (Commander Boyle), she operated on the North Atlantic "run" until the Spring of 1944. During this period, HAVELOCK with the corvettes PIMPERNEL and GODETIA had the misfortune to be the escort for the tanker convoy TM1 of nine vessels from Trinidad to Gibraltar, which lost seven vessels to U-boat attack between 3-12.1.43. The two surviving tankers arrived at Gibraltar on 14.1.43.

Between 16-20.3.43, HAVELOCK participated in one of the largest convoy battles of the war, when she was escorting the convoy SC122 with the frigate SWALE and the corvettes GODETIA, PIMPERNEL, BUTTERCUP, LAVENDER and SAXIFRAGE and the attached destroyer LEAMINGTON. Six U-boats were driven off and only three ships of the convoy lost.

HAVELOCK is seen whilst operating with the 9th D.F. on Atlantic escort duties in the winter of 1940/41.
(National Maritime Museum N31704)

During 6-7.43, HAVELOCK participated in operations in the Bay of Biscay and between 30.9.-8.10.43, and with her group (B5) escorted the ships carrying the equipment of No. 247 Group of the Royal Air Force from the U.K. to the Azores to establish new airfields on the islands.

The "NEPTUNE" Operation was covered by HAVELOCK as part of Support Group 14 until 7.44, when she started a refit at Liverpool which was completed during 9.44. She then returned to the 14th Escort Group until the war's end, but was, however, laid-up at Portsmouth between 2-4.45 because of defects. HAVELOCK's swan song was with her sister HESPERUS and aircraft of 201 Squadron of the R.A.F. when she sank U242 north-west of Anglesey on 30.4.45 by Hedgehog attack. She reduced to target service during 6.45, duties which she undertook for the next year. She was approved for scrapping on 18.2.46 and reduced to Category 'C' Reserve on 2.8.46. She was finally handed over to BISCO on 31.10.46 and immediately broken-up at Ward's Inverkeithing Yard.

HESPERUS (ex HEARTY, ex JURUENA) (H57)

Launched on 1.8.39 as the Brazilian JURUENA, she was requisitioned on 5.9.39 and renamed HEARTY and commissioned as such on 15.1.40. She was, however, renamed HESPERUS on 27.2.40.

She worked up at Portland and then joined the 9th D.F. of the Western Approaches Command. She had been hurried to sea and consequently suffered from leaking decks, had no form of gunnery control and an inadequate anti-aircraft armament as well as an ineffective gyro-compass. However her anti-submarine armament was massive by 1940 standards, because of the large number of depth charges carried.

HESPERUS undertook escort duties around Scapa Flow before being engaged in the Norwegian campaign. On 12.4.40 HESPERUS and HAVANT arrived at Thorshavn in the Faroes to report on the situation before the cruiser SUFFOLK with Royal Marines provided the occupying force. Three days later HESPERUS supported a troop landing at Mo, Norway, under severe air attack. However on 23.4.40 she was damaged by two near misses off Narvik and had to return to Dundee for repairs, which took a month. During these repairs her aft torpedo tubes were removed and a 3" high angle gun substituted.

HESPERUS then undertook escort duties for the next eight months (during 11.40 the 9th D.F. was designated the 9th Escort Group). In 1.41 HESPERUS with HURRICANE received weather damage whilst riding out a freak tropical storm. After damage repairs HESPERUS continued her Atlantic convoy duties until 4.41. She was then loaned to Force 'H' and formed part of the escort of the 'TIGER' convoy operation to Malta on 19.5.41 and Operation 'TRACER' between 13-15.6.41. Admiral Somerville ordered that as her anti-aircraft armament was so inadequate, HESPERUS should not serve on such operations again unless her armament had been improved. A short refit at Liverpool then followed when H/F D/F, and voice radio was fitted but not director control. She then joined the newly formed Newfoundland Escort Force on 7.7.41. During 8.41 she escorted the battleship PRINCE OF WALES carrying the Prime Minister to the Atlantic Charter meeting with President Roosevelt at Placentia Bay. HESPERUS subsequently received heavy weather damage which was repaired alongside a depot ship at Iceland, before proceeding to Immingham for permanent repairs, which included the fitting of director gear. She then returned to the 9th D.F. before being loaned to Force 'H' during 12.41.

Whilst with Force 'H', HESPERUS sank the first of the four U-boats she was to despatch in the next 16 months. On 15.1.42, whilst escorting convoy SL97 out of Gibraltar, she sank U93 by depth charges, gunfire and ramming. HESPERUS was flooded forward of station 14 and the whole of her starboard side was buckled including the bilge keel, with additional damage to her starboard propeller. After temporary repairs at Gibraltar HESPERUS left on 24.1.42 as escort for the troopship ALMANZORA to Falmouth. Repairs were completed there between 9.2-4.42 and radar and two 20mm guns were added. HESPERUS then became leader of Escort Group B2 on the dispersal of the 9th Escort Group.

HESPERUS then undertook escort duties for the next eight months. On 26.12.42 she and the destroyer VANESSA when escorting convoy HX219 attacked the U357 by depth charges. HESPERUS applied the coup de grace by ramming the submarine and rescuing survivors. HESPERUS again damaged her bow and the ship's bottom was ripped for nearly a quarter of her length and her speed reduced to 15 knots. Repairs undertaken at Liverpool were not completed until 17.3.43.

HESPERUS rejoined her group in time for the crucial convoy actions of 4-5.43 and quickly gained two more submarine successes; sinking the U191 without survivors with her newly fitted Hedgehog equipment on 23.4.43 whilst escorting ONS4. Nineteen days later she sank U186, also without survivors, whilst protecting SC129.

She remained on these arduous duties for the remainder of 1943 before refitting between 1.44-29.3.44.

HESPERUS on 9.9.44 as a member of the 19th Escort Group. Note the many war modifications.

HESPERUS then joined the 19th Escort Group, transferring to the 14th Escort Group during 1.45. She then operated from Portsmouth until the end of the war. On 30.4.45 she aided the sinking of U242 by her sister HAVELOCK. On 13.5.45 the 14th Escort Group with HESPERUS as senior ship escorted surrendered U-boats from Loch Alsh to Lough Foyle. A fortnight later HESPERUS and HAVELOCK escorted the liner ANDES carrying the Norwegian Government back to Oslo, staying there to 1.6.45.

On 11.6.45 HESPERUS was allocated for service as a Fleet Air Arm target ship and performed these duties for the next year. During 5.46 she started to reduce to Category 'C' reserve at Rosyth, having been approved for scrapping on 18.2.46. By 1.11.46 HESPERUS was confirmed as being in Category 'C' Reserve. However, by 26.11.46 she was lying at the yard of G. W. Brunton at Grangemouth for demolition, but was not officially handed over to BISCO until 17.5.47 when scrapping commenced.

HIGHLANDER (ex JAGUARIBE) (H44)

HIGHLANDER commissoned for trials on 18.3.40 and after a brief work-up at Portland she commenced escort duties with the 9th D.F. on 11.4.40. Six days later she went ashore in the Shetlands and suffered damage to one engine, her ASDIC, and her bottom. She repaired at the yard of Brigham and Cowan at Hull between 20.4-19.5.40.

HIGHLANDER then joined the destroyers ASHANTI and BULLDOG as escort for the battle-cruiser RENOWN on passage to Scapa where they arrived on 24.5.40. She did not remain long with the Home Fleet as on 12.6.40 she was ordered to join the Western Approaches station. On her passage south she picked up survivors from the torpedoed A. M. C. SCOTSTOUN and landed them at Greenock. HIGHLANDER was then involved in the evacuation of the B.E.F. from St. Nazaire on 16.6.40 and from St. Jean de Luz on 25.6.40.

HIGHLANDER fitting out at Thornycroft's, with HEARTY (later HESPERUS) alongside, some time after HIGHLANDER'S launch on 16.10.39.
(National Maritime Museum M1899)

She then returned to her escort duties with the 9th D.F. until 1.41, with absences for the fitting of a 3" H.A. mounting at Plymouth between 16-24.7.40, a convoy operation to Iceland during 10.40 and acting as escort for the Canadian destroyer SAGUENAY torpedoed and damaged on 1.12.40 and escorted into Barrow for repairs. A few weeks earlier HIGHLANDER and HARVESTER had sunk U32 in the North Atlantic on 30.10.40.

On 30.1.41 she arrived at Hawthorn Leslie's yard at Hebburn on Tyne for a refit, which was completed on 23.3.41. Three days later she sailed for Gibraltar to join the South Atlantic Command after a period of service with Force 'H' at Gibraltar. HIGHLANDER remained at Freetown with the 18th D.F. until 8.41. She then returned to the U.K. and as an unallocated vessel was based at Londonderry for the remainder of 1941 before joining the 28th Escort Group until 2.42.

On 2.2.42, HIGHLANDER was taken in hand by Green & Silley Weir at Tilbury to refit and rearm, which was completed on 18.3.42. She then became leader of Escort Group B4 and operated on the Atlantic for the next eight months. This arduous work meant that HIGHLANDER had to refit once more on the Humber between 22.11.42 and 6.1.43. She then resumed her duties with her Group and was plunged into the epic convoy action with HX229 when she was missed by two complete salvoes of torpedoes fired by U441 and U608.

Between 28.12.43 and 12.4.44 she refitted at Troon, work which was extended by a collision with a tug. After working up at Tobermory she rejoined her group until 9.44. In that month she joined Escort Group B2 on long range escort duties to St. John's Newfoundland until 4.45.

On 15.4.45 the corvettes MORPETH CASTLE and PRIMROSE were despatched to assist HIGHLANDER which had reported that she had suffered severe ice damage in position 45°48' N 49°05' W. The C.O. of HIGHLANDER reported:

"Request assistance as A/S instrument space and possibly other parts of forward bottom plates have carried away. I am steering stern first to minimise damage."

The tug TENACITY effected a rendezvous with HIGHLANDER and towed her into the Canadian base at the Bay of Bulls, Newfoundland, escorted by the Canadian corvette MEDICINE HAT which had joined.

HIGHLANDER on the marine railway at Bay of Bulls. Four views by courtesy of Arthur D. Rouse, who was serving on board her at the time.

Repairs at Bay of Bulls and later at St. John's were completed between 17.4-24.7.45. HIGHLANDER returned to Portsmouth on 29.7.45 where she remained for the next month. She then made passage to Rosyth where she completed her last few months service as a target. She was ordered to reduce to Category 'C' Reserve on 19.1.46. A month later on 19.2.46 she was approved for scrap. She was in Category 'C' Reserve on 2.8.46 and was reportedly handed over to BISCO on 27.5.46. She arrived at Charlestown about 2.47 and was then transferred to Metal Industries' Rosyth yard for demolition about 5.47.

HURRICANE on completion in 6.40. She carries the designed armament of 3 4.7" guns.
(Imperial War Museum FL494)

HURRICANE (ex-JAPARUA) (H06)

After working-up, HURRICANE joined her four sisters (HAVANT already having been lost) as part of the specialist A/S 9th D.F. However, the 9th D.F. was attached to the Home Fleet in the summer of 1940 until losses had been made good.

HURRICANE then spent until 5.41 on anti-submarine escort duties with the 9th D.F. The highlights of this portion of her career were the rescuing of survivors from the British liner CITY OF BENARES and the freighter MARINA, both torpedoed by U48 in position 56°46'N 21°15'W on 17.9.40. She also rescued no less than 451 survivors from the torpedoed steamer CITY OF NAGPUR. The survivors were taken to Greenock and landed on 1.5.41. Previously, on 22.2.41, HURRICANE and the corvette PERIWINKLE had unsuccessfully attacked a submarine whilst protecting convoy OB287.

On the night of 7/8.5.41, HURRICANE and the destroyer VISCOUNT were berthed on the outside of three groups of three escort vessels in Gladstone Dock, Liverpool. At 01.05 hours VISCOUNT was damaged by a bomb that exploded alongside and all ammunition, depth charges and torpedoes were transferred to the destroyer VANQUISHER. However at 03.00 hours, HURRICANE received a direct hit on the upperdeck aft at about station 132 on the port side. The bomb passed through the ship's side and exploded below the ship. A hole was blown in the outer bottom plating and the internal structure in the way of the explosion was wrecked. Her engine room and oil fuel tanks were flooded. The ship's after end commenced to sag and caused a deep corrugation in the ship's side in the vicinity of the explosion. The sloop ENCHANTRESS carried out pumping operations. In order to stop the possible blocking of other vessels, both HURRICANE and VISCOUNT were immediately moved to Bidston Dock as a precautionary measure. The Constructors' report noted that had the vessel been at sea and had received similar damage under heavy weather conditions, she would probably have been lost. There were no fatal casualties but a few members of the crew were treated for minor injuries.

HURRICANE's repairs and subsequent work up were not completed until 1.42 when she became leader of Escort Group B1 until her loss some twenty three months later. HURRICANE operated on the Atlantic run for the whole of this period, except for an anti U-Boat sweep in the Bay of Biscay with the Canadian destroyers ATHABASKAN and IROQUOIS and the Polish destroyer ORKAN.

On 24.12.43, she was ordered to assist the American Task Group 21.4 in a position approximately 420 miles NNE of the Azores. The U.S. destroyer LEARY had already been lost and the U.S.S.SCHENK attacked no less than five times. Shortly after HURRICANE had joined the task group she was torpedoed by U415 at 20.01 hours the same evening. The T5 'Gnat' torpedo blew off 30ft of HURRICANE's stern and although she had a 10° list to port and was unable to steam, she was in no danger of sinking. During the night, all heavy weights such as depth charges and torpedoes were jettisoned to lighten the ship and reduce her list. However at 13.01 hours the next day 25.12.43, in accordance with the C.in C's direct orders, she was torpedoed by the destroyer WATCHMAN after all hatches, flood valves, and sea cocks had been opened and sank in position 45°10'N 22°05'W. One rating was killed, two were missing and 9 wounded in the initial explosion.

THE TURKISH I's

Four destroyers of a slightly modified "I" class design which had been ordered by the Turks were under construction at the commencement of the war. However, for diplomatic reasons two of the order—SULTAN HISAR and DEMIR HISAR, both being built at Denny's of Dumbarton were completed for the Turks in 1942. The other vessels—MUAVENET and GAYRET, being built by Vickers-Armstrongs at Barrow were requisitioned by the Royal Navy and completed for the Royal Navy as INCONSTANT and ITHURIEL.

	LAID DOWN	LAUNCHED	COMMISSIONED
INCONSTANT	24.5.39	24. 2.41	24.1.42
ITHURIEL	24.5.39	15.12.40	3.3.42

INCONSTANT weather stained but little altered except for the bridge.

INCONSTANT (EX MUAVENET) (H49)

INCONSTANT was taken over by the Royal Navy on loan on 14.11.41 and after commissioning, worked up at Scapa and operated as an unattached vessel in Arctic waters in 3.42. Almost immediately she escorted a convoy around the Cape via Ponta Delgada (29.3), Freetown (4.4) and Capetown (19.4). On 5.5.42, she formed part of the covering force for the landing at Diego Suarez. INCONSTANT then proceeded to the Mediterranean and participated in the unsuccessful 'VIGOROUS' convoy operation from Alexandria to Malta (16.6-21.6.42). INCONSTANT then returned to the Eastern Fleet's 12th D.F. based at Kilindini, where she operated for the remainder of 1942. She participated in the occupation of Madagascar on 10.9.42.

INCONSTANT returned to the U.K. via the Cape and Freetown arriving at Plymouth on 25.2.43. She refitted at Devonport, which included her conversion to an anti-submarine destroyer, between 25.2-1.5.43. She then joined the 8th Escort Group of Western Approaches Command between 7.43-6.44. INCONSTANT had been one of the reinforcements sent to the Mediteranean for Operation 'HUSKY'—sinking U409 on 12.7.43 between Algiers and Bougie. Highlights of the next year included the escort of the convoy transporting air units establishing a base in the Azores. During 11.43, she protected a convoy to Murmansk, where she arrived on the 24th of that month. On her return to the U.K., she was under repair on the Clyde between 9.12.43-12.1.44. She then participated in escort duties in Icelandic and Arctic waters for the next four months, before being allocated to the 14th Escort Group for invasion duties between 4-8.44.

Her duties in the Channel complete, she refitted at Liverpool during 8.44 before briefly re-joining the 14th Escort Group. A further refit followed at the Blackwall yard of Green & Silley Weir between 6.10.44-22.12.44. The refit was completed at Devonport and she then served in the Irish Sea and Channel with the 14th Escort Group 1.45-7.45, based at Liverpool. She was under repair at Liverpool on V.E. day.

A couple of months were then spent operating out of Plymouth. On 18.9.45 she started a refit at Devonport to fit her for Turkish service, which was completed on 27.1.46. Official approval for her transfer to the Turks had been given on 3.10.45. She made passage to Istanbul, where after almost exactly four years service with the Royal Navy, she was subsequently handed over to the Turkish authorities at 11.00 hours on 9.3.46 as MUAVENET and served with the Turkish Navy until discarded in 1960.

ITHURIEL served for barely eight months before becoming a constructive total loss at Bone on 28.11.42. She was used as a base ship until 1945.

ITHURIEL (ex GAYRET) (H05)

On completion under low priority GAYRET was loaned to the Royal Navy as ITHURIEL for the duration of the war.

After a brief work up at Greenock, ITHURIEL sailed on 14.3.42 to join the 13th D.F. at Gibraltar on attachment. (She had originally been allocated to the Eastern Fleet, but never actually arrived on station). ITHURIEL was to be kept busy during the following months, participating in a flying-off operation with the aircraft carriers EAGLE and U.S.S. WASP during 5.42, and the ill fated 'HARPOON' convoy to Malta during 6.42, where she was slightly damaged by a bomb on 21.6.42 and was under repair for a week. She also participated in the 'PEDESTAL' convoy to Malta two months later.

On 12.8.42, whilst participating in this operation, ITHURIEL rammed and sank the Italian submarine COBALTO off Bizerta — three officers and 38 ratings from her crew were rescued. She was aided by the destroyer PATHFINDER. ITHURIEL returned to Gibraltar on 14.8.42 where temporary repairs were undertaken before she left 10 days later for Liverpool and Portsmouth. Action damage repairs and a refit were undertaken at Portsmouth Dockyard between 30.8-24.10.42. ITHURIEL was then re-allocated to Force 'H' and immediately participated in the 'TORCH' Landings.

On the night of 27-28.11.42, she was acting as emergency destroyer with steam available and was berthed at the North Quay of the large basin at Bone. At 00.50 hours on 28.11.42, two 500kg bombs fell alongside the ship — the first exploded below the aft magazine and shell room whilst the second bomb exploded under the ship abreast the engine room. ITHURIEL was severely shaken by the explosions with all lights failing and the engine room and aft magazine flooding as did the spirit room.

The destroyer QUENTIN assisted in pumping operations and towed ITHURIEL to No 6 quay where pumping operations continued as flooding remained unchecked. The ship was lightened with the removal of X and Y guns and attempts to remove fuel oil were made by the oiler BROWN RANGER.

It was ascertained that the ship's back was broken in two places between stations 100/102 and 151/154 and the stern had developed an ever increasing sag. The ship was beached on 29.11.42 with her bottom plating split and torn. This decision was made on the grounds that the ship could not be kept afloat indefinitely by pumping, she occupied a valuable berth and the general lack of salvage materials available. Temporary repairs were, however, undertaken.

On 27.2.43 she was towed to Algiers, where her damage was assessed and the decision made not to repair her. After being laid up at Algiers for several months, she was finally towed to Gibraltar on 18.8.43 where she was placed on a care and maintenance basis and used as a base ship for anti-submarine training for the next two years.

On 1.8.45, ITHURIEL, escorted by the trawler WINDERMERE, left Gibraltar in tow of the tug PROSPEROUS for Plymouth, where she arrived a week later. The tow was then continued to Bo'ness where ITHURIEL arrived on 13.8.45 for demolition by P&W MacLellan Ltd. In 1946, the destroyer ORIBI was ceded to the Turkish Navy as a replacement for ITHURIEL and given the name GAYRET.

APPENDIX I
DEPLOYMENT

On 3.9.39 the A/I's had started to be replaced in the Home and Mediterranean Fleets by the new "Tribal", J and K Class Destroyers. ECHO was to remain in the 7th Destroyer Flotilla a few more days until relieved by JAGUAR. The 81 destroyers in service were distributed as follows:

Home Waters : (37)
Home Fleet	:	8th Destroyer Flotilla (D.F.) — 9 F Class
Channel Force	:	18th D.F. — KEMPENFELT, ACASTA, ACHERON, ANTELOPE, ARDENT (5)
Humber Force	:	7th D.F. — ECHO with 7 J Class (1)
Other Stations	:	12th D.F. — EXMOUTH, ELECTRA, ECLIPSE, ENCOUNTER, ESCAPADE, ESCORT (6)
	:	18th D.F. — ARROW, ANTHONY, ACHATES, AMAZON (4)
Dover	:	19th D.F. — CODRINGTON, KEITH, BASILISK, BOREAS, BEAGLE, BLANCHE, BRILLIANT, BOADICEA, BRAZEN (9)

Refitting at Portsmouth: Minelayers ESK, EXPRESS, AMBUSCADE (3)

Mediterranean Fleet (24)
1st D.F. — 9 G Class
2nd D.F. — HARDY, HERO, HEREWARD, HASTY, HOSTILE (5)
3rd D.F. — 9 I Class
Attendant destroyer to the aircraft carrier GLORIOUS: BULLDOG
Gibraltar: 13th D.F. — ACTIVE

China (9)
21st D.F. — 9 D Class (on passage to the Mediterranean)

South Atlantic (4)
2nd D.F. — HYPERION, HUNTER, HOTSPUR, HAVOCK

Canada (6)
Halifax : SKEENA, SAGUENAY, Vancouver : ST. LAURENT, Esquimalt : OTTAWA, RESTIGOUCHE. On passage from Vancouver to Halifax : FRASER

During the first days of the war six similar vessels building for the Brazilians were requisitioned whilst under construction and were completed between 1939/40. Later two destroyers building for Turkey were also requisitioned.

The first months of the war saw the concentration of the A/I's in Home Waters by 11.39 excepting the D's held in the Mediterranean for repairs and work-up, those H's still on anti-raider duty in the South Atlantic and the Canadian vessels retained in Canadian and Caribbean waters. GARLAND was still under repair at Malta. By the end of the year BLANCHE and GIPSY had been lost by mines and DUCHESS by collision, but the first of the Brazilian vessels — HAVANT — had commissioned.

KEMPENFELT had also transferred to the Canadians as ASSINIBOINE to complete their half flotilla.

1940 was to see the A/I's heavily engaged in French and Norwegian waters and in the Mediterranean after Italy joined the war during 6.40. Considerable reinforcements of destroyers were sent to the Mediterranean in 5.40 prior to Italy's declaration of war. No fewer than 25 A/I's were lost during the year — CODRINGTON, KEITH, BASILISK, BRAZEN, DELIGHT, GRENADE, and HAVANT by air attack, ACHERON, ESK, GRENVILLE, HOSTILE, HYPERION and IVANHOE by mine, ACASTA, ARDENT, HARDY, GLOWWORM, HUNTER and GRAFTON by surface action, DARING, EXMOUTH, ESCORT by submarine attack and RCN FRASER, MARGAREE (ex DIANA) and IMOGEN by collision. These losses were partly compensated by the completion of the last 5 Brazilian H's. GARLAND was transferred to the Polish Navy.

The Pink List of 30.12.40 gave the distribution of the A/Is as follows:

Home Waters (32)
Home Fleet	:	3rd D.F. : ECHO, INGLEFIELD
		4th D.F. : ESCAPADE, ECLIPSE, ELECTRA, BULLDOG, BEAGLE, BRILLIANT, FAME (9)
The Nore	:	20th D.F. (M/L) : INTREPID, ICARUS, IMPULSIVE and EXPRESS (under repair) (4)

Portsmouth : Refitting : BOREAS, BOADICEA (2)

Western Approaches : 3rd Escort Group (E.G.) : AMBUSCADE, AMAZON, ARROW
4th E.G. ACTIVE, ANTELOPE, ANTHONY, ACHATES (repairs)
9th E.G. 5 Brazilian "H"
10th E.G. GARLAND, OTTAWA, SAGUENAY, SKEENA, ST. LAURENT (17)

Halifax Escort Force (2): ASSINIBOINE, RESTIGOUCHE

Gibraltar (11): 13th D.F.: ISIS, DUNCAN, ENCOUNTER
8th D.F.: FAULKNOR, FEARLESS, FIREDRAKE, FORESTER, FORESIGHT, FORTUNE, FOXHOUND, FURY

Mediterranean (14)
2nd D.F. : ILEX, HERO, HASTY, HEREWARD, HAVOCK
16th D.F. : DAINTY, DIAMOND, DEFENDER, DECOY
14th D.F. : GREYHOUND, GALLANT, GRIFFIN, HOTSPUR, IMPERIAL

TOTAL: 59

During 1941 the A/I's were not only deeply involved in Home Waters and the Mediterranean, but increasing numbers were being used on North Atlantic convoy duties. 7.12.41 was also to see the opening of the Pacific War, with A/I destroyers being involved from the very first day. During the year, the vessels' lack of a dual purpose armament contributed to the loss of no fewer than seven of the group, with GREYHOUND, HEREWARD, IMPERIAL, DIAMOND, DAINTY, DEFENDER and FEARLESS being sunk by air-attack in the Mediterranean. GALLANT was badly damaged by a mine, but was only declared a constructive total loss after further damage was received whilst under repair at Malta.

The situation on 2.1.42 is given by the Pink List of that date:

Home Waters (24)
Home Fleet : 3rd D.F.: INGLEFIELD, ECHO, ESCAPADE, ICARUS, IMPULSIVE, ECLIPSE, INTREPID
8th D.F.: FAULKNOR, FORESTER, FORESIGHT, FURY (11)

Western Approaches
3rd E.G. : AMBUSCADE, BULLDOG, AMAZON
4th E.G. : BOADICEA, BEAGLE
23rd E.G. : ACHATES, ANTELOPE
24th E.G. : FAME
25th E.G. : GARLAND
28th E.G. : HARVESTER, HIGHLANDER, HURRICANE, HAVELOCK (13)

Gibraltar (5) : 13th E.G.: DUNCAN
Force 'H': ANTHONY, HESPERUS, ACTIVE, FORTUNE

Mediterranean (8) : 2nd D.F. : HASTY, HOTSPUR, HERO, DECOY, HAVOCK, GRIFFIN
Unallocated : ARROW, FOXHOUND (under repair)

Eastern Fleet (3) : ELECTRA, ENCOUNTER, EXPRESS

South Atlantic (2) : 18th D.F. : BOREAS, BRILLIANT

Newfoundland (6) : ASSINIBOINE, OTTAWA, SAGUENAY, SKEENA, ST. LAURENT, RESTIGOUCHE

U.S.A. (1) : FIREDRAKE (refitting)

Bombay (1) : ILEX (refitting)

Singapore (1) : ISIS (repairing)

TOTAL: 51

1942 was to see the completion of the final vessels of the group — the ITHURIEL and INCONSTANT, but a further nine vessels were lost — FORESIGHT, ITHURIEL, HASTY and HAVOCK in the Mediterranean, ACHATES in the Arctic, OTTAWA (I) and FIREDRAKE in the North Atlantic and ELECTRA and ENCOUNTER during the disastrous Java campaign. The process of converting the destroyers to A/S vessels continued apace and by the end of the year only the Home Fleet's 8th D.F. was purely on Fleet duties. Many vessels had transferred from the Mediterranean to the Eastern Fleet, but by the end of the year, the comparative quiet of that theatre, the availability of newer vessels and the need for refit meant that a steady stream of vessels returned to the U.K.

The distribution of the survivors on 25.12.42 was as follows:

Home Waters (29)
Home Fleet:	:	8th D.F. FAULKNOR, ECHO, ECLIPSE, FORESTER, FURY, ICARUS, IMPULSIVE, INGLEFIELD, INTREPID (9)
Western Approaches	:	Londonderry : B1 — HURRICANE, B4 HIGHLANDER (2) Liverpool : B2 HESPERUS, B5 HAVELOCK, B6 FAME (3)
Special Escort	:	ACTIVE, ANTHONY, ARROW, DECOY, DUNCAN, GRIFFIN, (All refitting) (6) Greenock: B3 GARLAND, HARVESTER, ESCAPADE (refit) (3)
	:	ACHATES*, AMAZON, BEAGLE, BOADICEA, BULLDOG (5) *Lost 31.12.42.
Aircraft Target Ship	:	AMBUSCADE

Mediterranean (3) 22nd D.F.: HERO
Unallocated: ILEX, ISIS

Gibraltar (2) : 13th D.F.: BOREAS, BRILLIANT (attached)

Eastern Fleet (5) : 2nd D.F.: HOTSPUR
12th D.F.: EXPRESS, FORTUNE, FOXHOUND, INCONSTANT

Newfoundland Command (5)
ASSINIBOINE, RESTIGOUCHE, SAGUENAY, SKEENA, ST. LAURENT
South Atlantic (1) : ANTELOPE (on "Torch" convoy Duties)
TOTAL: 45

The major developments during 1943 were the transfer of the 8th D.F. to the Mediterranean, the continuation of the programme of anti-submarine escort conversions and the turning over of vessels — DECOY (KOOTENAY), EXPRESS (GATINEAU), FORTUNE (SASKATCHEWAN), GRIFFIN (OTTAWA (II)), HERO (CHAUDIERE) to the Canadians. Five of the class became losses during the year — INTREPID by air attack and ECLIPSE by mine in the Aegean, HARVESTER and HURRICANE to submarine attack in the Atlantic, whilst ARROW was declared a constructive total loss after the explosion of an ammunition ship. The vessels were distributed as follows on 31.12.43:

U.K. (26)
(2) Working up at Scapa : IMPULSIVE, CHAUDIERE
(3) Refitting : QU'APPELLE (ex FOXHOUND), DUNCAN, ESCAPADE
(1) Trial Duties : AMBUSCADE
(2) Under Care and Maintenance : AMAZON, BULLDOG
(18) Western Approaches : Londonderry : B.7 HIGHLANDER (Refit), C1 ASSINIBOINE, ST. LAURENT, FORESTER, C2 — ICARUS, GATINEAU, C3—SASKATCHEWAN, SKEENA, C4—HOTSPUR, RESTIGOUCHE, C5—KOOTENAY, OTTAWA (12)
Liverpool — B2—HESPERUS, B5—HAVELOCK, B6—FAME (3)
Greenock — 8th E.G. BEAGLE, INCONSTANT, BOADICEA
Gibraltar (6) — 13th D.F. ACTIVE, ANTELOPE, ANTHONY, BRILLIANT, ISIS, BOREAS
Mediterranean (5) : FAULKNOR, ECHO, FURY, ILEX, INGLEFIELD
Canada (1) : SAGUENAY (Base ship)
South Atlantic (1) : GARLAND.
TOTAL: 39

Operation "OVERLORD" resulted in a concentration of the A/I's in Home Waters. A further five vessels of the class were lost during the year — INGLEFIELD by glider bomb attack off Anzio, BOADICEA, FURY and ISIS during operations in the Channel supporting "OVERLORD", whilst SKEENA was wrecked in Iceland during 10.44. This year also marked the end of the A/I's on fleet duties and saw FOXHOUND transferred to Canada as QU'APPELLE and BOREAS and ECHO being loaned to the Greeks as SALAMIS and NAVARINON. An increasing number of vessels were also laid up as the years of arduous service and minimal repairs took their toll.
The Pink List of 29.12.44 listed the following vessels:

Active Vessels (30)
Portsmouth (4)   : First D.F.: ACTIVE, ANTHONY, BRILLIANT, GARLAND
Plymouth (2)   : 8th D.F.: FAULKNOR, IMPULSIVE

Londonderry (8)	:	11th E.G.: GATINEAU, KOOTENAY, RESTIGOUCHE, CHAUDIERE*, OTTAWA (II)*, QU'APPELLE*, ST LAURENT*, SASKATCHEWAN* *Refitting
Liverpool (10)	:	B2 : HIGHLANDER, 19th E.G. : HESPERUS 14th E.G.: DUNCAN, HAVELOCK, ICARUS, INCONSTANT, ASSINIBOINE, FAME, FORESTER, HOTSPUR
Greenock (3)	:	7th E.G.: BULLDOG, 8th E.G. BEAGLE, Unallocated : ESCAPADE
Mediterranean (3)	:	SALAMIS, NAVARINON, ILEX (Repairing)

Reserve and Training (5)
ANTELOPE (Tyne), AMBUSCADE (Training), ILEX (Bizerta), AMAZON (Trials), SAGUENAY (H.M.C.S. CORNWALLIS)
TOTAL: 35

At the war's end the distribution of the vessels was as follows:

ACTIVE

Plymouth (6)	:	8th D.F. FAULKNOR, GARLAND, IMPULSIVE Unattached : BEAGLE, BULLDOG, KOOTENAY
Western Approaches (7)	:	8th E.G. ESCAPADE, 11th E.G. SASKATCHEWAN, 14th E.G. ICARUS, HAVELOCK, HESPERUS, INCONSTANT, ASSINIBOINE
Irish Sea (3)	:	HOTSPUR, DUNCAN, GATINEAU
Canadian Waters (5)	:	RESTIGOUCHE, ST. LAURENT, OTTAWA (II), HIGHLANDER (non operational), SAGUENAY (Base ship)
Mediterranean (3)	:	3rd D.F. ACTIVE plus Greek NAVARINON, SALAMIS
Refitting (4)	:	FAME (Leith), FORESTER (Liverpool), QU'APPELLE (Halifax), CHAUDIERE (Sydney)
Converting to Target Ships (2)	:	ANTHONY, BRILLIANT
Test Vessel (1)	:	AMAZON
Reserve (2)	:	ILEX (Malta), ANTELOPE (Tyne)
Training Vessel (1)	:	AMBUSCADE

TOTAL: 34

The majority of the A/I's had seen strenuous war service and it is not surprising that the surviving vessels were quickly paid off and scrapped. By the end of 1946 only FAME and HOTSPUR were active as training vessels, with ACTIVE and AMBUSCADE being used for shock-trials and with AMAZON, ANTHONY, BRILLIANT, ESCAPADE and GARLAND in various categories of Reserve. All the remaining vessels, including the whole of the Canadian force had been sold, except for INCONSTANT which had been returned to Turkey. She was the only one of the eight requisitioned vessels to see service with the navies that had ordered them.

INCONSTANT at Malta following her delivery to the Turkish Navy as MUAVENET. Note the identifying letter painted on the hull.

APPENDIX II

ARMAMENT CHANGES TO THE A/I CLASS DESTROYERS ON CONVERSION TO ANTI-SUBMARINE ESCORTS

By 9.39, the A and B's had been relegated to what were then regarded as second line duties — those of convoy escort and local patrol. The war was to bring about profound changes in not only these vessels but all the A/I's.

The ex-Brazilian vessels were assigned to escort duties from their completion and were only utilised as fleet escorts in times of extremis. The availability of newly constructed fleet destroyers in larger numbers from the spring of 1942 meant that the A/I's were increasingly used as convoy escorts.

The fitting of modified armament and sensor equipment was, however, largely dependent on the vessel being in dockyard hands at the time the equipment was available. Complete refits were rare and it seems were only undertaken when the vessel was repairing serious damage (e.g. FAME, FIREDRAKE) or in need of a long refit on their return from service with the Eastern Fleet (e.g. DUNCAN, GRIFFIN and HOTSPUR). Gradually two types of conversion evolved:

(a) **THE THREE GUN ARRANGEMENT** consisted of:
    3 4.7" (with 160 rounds of SAP, 70 rounds of high explosive per gun); No. 4 mounting removed.
    3 2 pounder (singles)
    6 20mm Oerlikon
    1 .303" Lewis gun (Stripped)
    2 PAC projectors

As far as can be ascertained the following vessels were converted to this armament arrangement:
ACHATES, ACTIVE, ANTELOPE, ANTHONY, ARROW, BOREAS, ASSINIBOINE, OTTAWA(I), DUNCAN, KOOTENAY, ECHO, ECLIPSE, GATINEAU, FAULKNOR, FAME, FORESTER, SASKATCHEWAN, QU'APPELLE, FURY, GARLAND, CHAUDIERE, HOTSPUR, ICARUS, INCONSTANT.

(b) **THE TWO GUN ARRANGEMENT** consisted of:
    2 4.7" (singles) (usually No. 2 and No. 3 mountings retained, but dependent on fitting of Hedgehog).
    2 2 pounder
    2 6 pounder (Hotchkiss)
    6 20mm Oerlikon (4 in HAVANTs)
    2 FAC projectors
    1 .303" Lewis gun (Stripped)

By 1944/45 the following vessels had completed to this arrangement:
AMAZON, AMBUSCADE (before use as a trials vessel), SAGUENAY, SKEENA, BEAGLE, BOADICEA, BULLDOG, BRILLIANT, ST. LAURENT, ESCAPADE, FIREDRAKE, OTTAWA (II) and the Brazilian 'H's.

As can be appreciated, the vessels lost between 9.39-12.41 had either not been altered or only had their anti-aircraft armaments supplemented.

The following vessels lost during 1942/44 had not been altered:
ELECTRA, ENCOUNTER, FORESIGHT, FURY, HASTY, HAVOCK, INGLEFIELD, INTREPID, ISIS and ITHURIEL. ILEX was never altered as she had been laid up since early 1944 and IMPULSIVE it seems was never materially altered. Her armament at the end of the war consisted of 4.47", 4 Oerlikons, 44 depth charges, 8 torpedo tubes and a 5 charge depth charge pattern.

ANTI-SUBMARINE WEAPONS

Originally designed with two depth charge throwers, one rail and 15 depth charges, later increased in the 'G' Class and subsequent vessels to 20 depth charges.

In the early days of the war 35 depth charges were being carried.

The number of depth charges was later increased to 60, then 125 and finally in some vessels to a total of 135. Changes were also made in the number of depth charges dropped in a pattern; following analysis undertaken by the Admiralty operational researchers, from five per pattern to seven and finally to ten per pattern.

The major change was the introduction of ahead throwing weapons, which made it harder for submarines to escape when ASDIC readings were lost as the attacking vessel passed directly over the submarine. The first of these weapons was Hedgehog which was fitted originally in lieu of No. 1 gun mounting, but because the "electrics" were frequently damaged by water, the system was moved to No. 2 position in some ships. AMBUSCADE was used as a trial vessel for the prototype SQUID mounting.

RADAR EQUIPMENT FITTED TO THE A/I's

At the beginning of the war, none of the A/I's were fitted with radar or R.D.F. (Radio Direction Finding) as it was then known. Naval radar had been developed by the Admiralty Signals Establishment between 1935-1938. The first set (Type 79Y) — an air-warning set — was fitted in the cruiser SHEFFIELD during 10.38, followed by another in the battleship RODNEY three months later.

By 1940 the need for similar air-warning, surface warning (W.S.) — to detect surfaced submarines and gunnery fire-control (G.A.) radars had already been recognised. The fitting of surface warning radars in destroyers commenced in 1941 and the fitting programme had been completed in surviving units by 1944.

The amount of equipment fitted varied considerably between vessels, the A/S vessels did not need G.A. sets, operational requirements limited the time available for fitting radars, and there were delays in producing sets. Installations were frequently carried out concurrently with repairs or during boiler-cleaning periods.

THE TYPES OF RADAR FITTED
(i) **Surface Warning Sets**

The Type 271 was the first centimetric radar, with a manually rotated aerial housed in a protective 'Lantern' radome above the bridge. Small ship detection range was six miles. Modifications of the type were the 271M, 271P, 271Q and 272/273 radars.

The Type 286M was a modified R.A.F ASV MkII, with a range of 5,000 yards for submarines and six miles for large ships. Developments were types 286P, 286Q, and the Type 290 was an improved 286P fitted with a ranging panel to provide the surface range of targets.

(ii) **Warning Aircraft/Warning Surface Combined (W.A./W.S)**

The Type 291, with a manually rotated aerial had a range of 30 miles and 10,000 feet for aircraft and eight/nine miles for surface ships. This set replaced the Type 286/289 sets. The Type 291M was a development of the Type 291 with a power driven aerial and a Plan Position Indicator. This set was not fitted until 1945 in destroyers.

(iii) **Gunnery Sets**

The Type 285 had Yagi (fishbone) type aerials mounted on a H.A. Director. The set provided accurate ranges to the fire control system up to 18,000 yards. The S.G.1 was a U.S.N. set with a power driven aerial. Range was 15-22 miles for surface warships and 10 miles for surfaced submarines.

(iv) **Identify Friend or Foe (I.F.F.) Equipment**

There were several types. The Type 242, used in conjunction with Type 271 and aerial mounted above it. Type 252 beacon was used in conjunction with Type 286, 290 and 291 radars. The Type 253 was an improved version.

Typical Installations by 1944 were as follows:

AMAZON, AMBUSCADE	271, 286M (later 290)
ACTIVE, ANTELOPE, ANTHONY, ARROW	271, 242 IFF, 286M (later 290)
ACHATES (At Loss)	271, 286M
BOADICEA, BOREAS, BRILLIANT, BULLDOG	271, 242 IFF, 290
BEAGLE	286M
ASSINIBOINE (RCN)	286M
DECOY	290
DUNCAN	271, 242, IFF, 286M (later 291)
ESCAPADE	271, 242 IFF, 291 (Unusually Type 277 was fitted in 1945)
FAME	271, 242 IFF, 252, 284. SG1 fitted on transfer to the Dominican Republic
FAULKNOR	285, 290, 242 IFF
HASTY (At loss)	286M
HOTSPUR	271, 242 IFF, 291 (Later) SG1 and 291 fitted on transfer to the Dominican Republic)
ICARUS	286M (later 290), 252 IFF

APPENDIX III

ARMAMENT FIT OF FAME AS AN ANTI-SUBMARINE DESTROYER ON TRANSFER
TO THE DOMINICAN REPUBLIC ON 18.12.48.

FAME incorporated all the wartime changes in armament and electronics suite to fit her for these duties. Her armament on transfer was listed as:

    3 4.7" in A, B and X positions
    4 Single bofors: 2 each on the signal deck and 2 on the searchlight platform
    1 Set of quadruple torpedo tubes (forward)
    Equipment for a 10 Depth Charge pattern
    4 depth charge throwers and 2 rails for 70 depth charges
    Type 291 and SG1 Radars
    Type 128 ASDIC

She shipped 20 tons of permanent ballast and her GM in light condition was 2.4 feet.

This armament statement gives an indication of how the original torpedo and gun armament had been superseded by anti-submarine, light anti-aircraft and modern electronics equipment.

INDEX

This Index gives the name and pennant number of each of the ships with which the book is concerned. It gives the number of the page on which each ship's history begins and also the numbers of pages on which photographs appear if not accompanying the history concerned. References in the Appendices are also included. The D flag superior changed to I in April 1940.

Name	Pennant	Pages	Name	Pennant	Pages
ACASTA	H09	17, 137	GIPSY	H63	95
ACHATES	H12	17, 137, 138, 139, 141, 142	GLOWWORM	H92	95
ACHERON	H45	19, 137	GRAFTON	H89	97
ACTIVE	H14	20, 137, 138, 139, 140, 141, 142	GRENADE	H86	98
AMAZON	D39, I39	F'piece, 11, 137, 138, 139, 140, 141, 142	GRENVILLE	H03	90
AMBUSCADE	D38, I38	12, 137, 138, 139, 140, 141, 142	GREYHOUND	H05	99, 138
ANTELOPE	H36	21, 137, 138, 139, 140, 141, 142	GRIFFIN	H31	100, 138, 139, 141
ANTHONY	H40	22, 137, 138, 139, 140, 141, 142	HANDY	H19	127
ARDENT	H41	23, 137	HARDY	H87	103, 137
ARROW	H42	24, 137, 138, 139, 141, 142	HARVESTER	H19	127, 138, 139
ASSINIBOINE	D18, I18	45, 137, 138, 139, 140, 141, 142	HASTY	H24	104, 137, 138, 141, 142
BASILISK	H11	32, 137	HAVANT	H32	129,
BEAGLE	H30	32, 137, 138, 139, 140, 141, 142	HAVELOCK	H88	129, 138, 139, 140
BLANCHE	H47	33, 137	HAVOCK	H43	105, 137, 138, 141
BOADICEA	H65	34, 137, 138, 139, 141, 142	HEARTY	H57	129
BOREAS	H77	36, 137, 138, 139, 141, 142	HEREWARD	H93	106, 137, 138
BRAZEN	H80	37, 137	HERO	H99	107, 137, 138, 139
BRILLIANT	H84	39, 137, 138, 139, 140, 141, 142	HESPERUS	H57	130, 138, 139, 140
BULLDOG	H91	40, 137, 138, 139, 140, 141, 142	HIGHLANDER	H44	131, 138, 139, 140
CHAUDIERE	H99	107, 139, 140, 141	HOSTILE	H55	109, 137
CODRINGTON	D65, I65	16, 137	HOTSPUR	H01	110, 137, 138, 139, 140, 141, 142
COMET	H00	46	HUNTER	H35	112, 137
CRESCENT	H48	48	HURRICANE	H06	134, 138, 139
CRUSADER	H60	48	HYPERION	H97	113, 137
CYGNET	H83	50	ICARUS	D03, I03	116, 138, 139, 140, 141, 142
DAINTY	H53	53, 138	ILEX	D61, I61	118, 138, 139, 140, 141
DARING	H16	54	IMOGEN	D44, I44	120
DECOY	H75	55, 138, 139, 142	IMPERIAL	D09, I09	120, 138
DEFENDER	H07	57, 138	IMPULSIVE	D11, I11	121, 138, 139, 140, 141
DELIGHT	H38	58	INCONSTANT	H49	135, 138, 139, 140, 141
DIAMOND	H22	58, 138	INGLEFIELD	D02, I02	115, 138, 139, 141
DIANA	H49	59	INTREPID	D10, I10	123, 138, 139, 141
DUCHESS	H64	60	ISIS	D87, I87	124, 138, 139, 141
DUNCAN	H99	52, 138, 139, 140, 141, 142	ITHURIEL	H05	136, 138, 141
ECHO	H23	64, 137, 138, 139, 141	IVANHOE	D16, I16	126
ECLIPSE	H08	66, 137, 138, 139, 141	JAGUARIBE	—	131
ELECTRA	H27	68, 137, 138, 141	JAPARUA	—	134
ENCOUNTER	H10	69, 137, 138, 141	JAVARY	—	129
ESCAPADE	H17	70, 137, 138, 139, 140, 141, 142, 144	JURUA	—	127
ESCORT	H66	71, 137	JURUENA	—	130
ESK	H15	72, 137	JUTAHY	—	129
EXMOUTH	H02	64, 137	KEITH	D06, I06	30, 137
EXPRESS	H61	73, 137, 138, 139	KEMPENFELT	D18, I18	45, 137
FAME	H78	77, 138, 139, 140, 141, 142	KOOTENAY	H75	55, 139, 140, 141
FAULKNOR	H62	76, 138, 139, 140, 141, 142	MARGAREE	H49	59
FEARLESS	H67	78, 138	MARNIX	—	94
FIREDRAKE	H79	79, 138, 141	MUAVENET	—	135, 140
FORESTER	H74	82, 138, 139, 140, 141	NAVARINON	H23	66, 139, 140
FORESIGHT	H68	80, 138, 141	OTTAWA(I)	H60	48, 137, 138, 141
FORTUNE	H70	83, 138, 139	OTTAWA(II)	H31	6, 100, 139, 140, 141
FOXHOUND	H69	84, 138, 139	QU'APPELLE	H69	85, 139, 140, 141
FRASER	H48	48, 137	RESTIGOUCHE	H00	46, 137, 138, 139, 140
FURY	H76	86, 138, 139, 141	SAGUENAY	D79, I79	26, 137, 138, 139, 141
GALLANT	H59	91, 138	ST. LAURENT	H83	50, 137, 138, 139, 140, 141
GARLAND	H37	92, 138, 139, 140, 141	SALAMIS	H77	36, 139, 140
GATINEAU	H61	73, 139, 140, 141	SASKATCHEWAN	H70	83, 139, 140, 141
GAYRET	—	136	SKEENA	D59, I59	28, 137, 138, 139, 141
GENERALISIMO	—	78	TRUJILLO	—	112

ACKNOWLEDGEMENTS

In preparing this book extensive use has been made of the official records held at the Public Record Office at Kew, in the National Maritime Museum archives at Woolwich and by the Naval Historical Branch of the Ministry of Defence (Navy) and I would like to make particular reference to the help afforded by David Brown, Head of the Naval Historical Branch and Mike McAloon also of that Branch and by David Hodge and Paul Kemp of the Photographic Departments of respectively, the National Maritime Museum and the Imperial War Museum. Their assistance was invaluable. I am also indebted to Michael Crowdy for his work in proof-reading and assisting in the presentation and production of this book.

PHOTOGRAPHS

Unless otherwise stated all photographs are from the World Ship Society's various collections. Other important sources are the Imperial War Museum and the National Maritime Museum and photographs that can be obtained from these sources are identified appropriately with the relevant negative number: enquiries for prints of them should be directed accordingly. Other photographs have been supplied by the Royal Netherlands Navy, Ken Royall, Strathclyde Regional Council Archive, Swan Hunter Ltd., V.S.E.L. and Roy Wilson. Photographs credited "W.S.P.L." are from negatives in the possession of the World Ship Society.

WORLD SHIP SOCIETY PUBLICATIONS

"AMAZON TO IVANHOE" is one of a number of comprehensively illustrated books published by the World Ship Society and dealing with maritime aspects of World War II.

Uniform with "AMAZON TO IVANHOE" are "THE HUNTS", "THE TOWNS" and "SLOOPS 1926-1946", which also give extensive constructional and performance details of these destroyer and escort types as well as the individual careers and fates of the ships concerned.

Other World Ship Society publications specifically relating to the Second World War and available in 1993 are "CONVERSION FOR WAR", which gives extensive information about modifications made to passenger liners and certain cargo vessels to fit them for naval duties, and "CONVOYS TO RUSSIA, 1941-1945", which provides a chronology of all these convoys with the names of every naval and merchant ship which took part, details of the attacks which took place, and of German and Allied successes. "TYPE 35 TORPEDO BOATS OF THE KRIEGSMARINE" gives details and war histories of these craft and "REGISTER OF TYPE VII U-BOATS" provides dates of order, launch, commissioning and fate of these numerous submarines together with specifications and armament details.

In addition, the World Ship Society has published the fleet histories of over sixty shipping companies — Danish, Dutch, German, Norwegian and Swedish, as well as British — and each of these gives details relating to the losses of individual ships during the war.

For details of the World Ship Society, its services and publications, please write to Department Al, 5 Grove Road, Preston, PR5 4AJ, England.

The badly damaged ESCAPADE following the explosion 20.9.43.